D0140568

STUDENT SOLUTIONS MANUAL FOR ZILL'S

A First Course in DIFFERENTIAL EQUATIONS

THE CLASSIC FIFTH EDITION

WARREN S. WRIGHT
Loyola Marymount University

CAROL D. WRIGHT

BROOKS/COLE
CENGAGE Learning™

Australia • Brazil • Japan • Korea • Mexico • Singapore • Spain • United Kingdom • United States

**Student Solutions Manual for Zill's A First
Course in Differential Equations,
The Classic Fifth Edition**
Warren S. Wright, Carol D. Wright

Assistant Editor: Carol Benedict

Editorial Assistant: Daniel Thiem

Marketing Manager: Karin Sandberg

Marketing Assistant: Laurie Davidson

Production Coordinator: Dorothy Bell

Cover Design: Roy R. Neuhaus

Cover Photo: The Stock Market/
M. Mastrorillo

Print Buyer: Micky Lawler

© 2001 Brooks/Cole, Cengage Learning

ALL RIGHTS RESERVED. No part of this work covered by the copyright
herein may be reproduced, transmitted, stored or used in any form or by
any means graphic, electronic, or mechanical, including but not limited to
photocopying, recording, scanning, digitizing, taping, Web distribution,
information networks, or information storage and retrieval systems, except
as permitted under Section 107 or 108 of the 1976 United States Copyright
Act, without the prior written permission of the publisher.

For product information and technology assistance, contact us at
Cengage Learning Customer & Sales Support, 1-800-354-9706

For permission to use material from this text or product,
submit all requests online at **cengage.com/permissions**
Further permissions questions can be emailed to
permissionrequest@cengage.com

ISBN-13: 978-0-534-38280-3

ISBN-10: 0-534-38280-0

Brooks/Cole
10 Davis Drive
Belmont, CA 94002-3098
USA

Cengage Learning is a leading provider of customized learning solutions with
office locations around the globe, including Singapore, the United Kingdom,
Australia, Mexico, Brazil, and Japan. Locate your local office at:
international.cengage.com/region

Cengage Learning products are represented in Canada by
Nelson Education, Ltd.

For your course and learning solutions, visit **academic.cengage.com**

Purchase any of our products at your local college store or at our preferred
online store **www.ichapters.com**

Printed in the United States of America
1 2 3 4 5   15 14 13 12 11

FD102

Table of Contents

1 Introduction to Differential Equations

<hr>

Exercises 1.1

3. First-order; nonlinear because of yy'.

6. Second-order; nonlinear because of $\sin y$.

9. Third-order; linear.

12. From $y = 8$ we obtain $y' = 0$, so that $y' + 4y = 0 + 4(8) = 32$.

15. From $y = 5\tan 5x$ we obtain $y' = 25\sec^2 5x$. Then

$$y' = 25\sec^2 5x = 25\left(1 + \tan^2 5x\right) = 25 + (5\tan 5x)^2 = 25 + y^2.$$

18. First write the differential equation in the form $2xy + \left(x^2 + 2y\right)y' = 0$. Implicitly differentiating $x^2 y + y^2 = c_1$ we obtain $2xy + \left(x^2 + 2y\right)y' = 0$.

21. Implicitly differentiating $y^2 = c_1\left(x + \dfrac{1}{4}c_1\right)$ we obtain $y' = c_1/2y$. Then

$$2xy' + y(y')^2 = \frac{c_1 x}{y} + \frac{c_1^2}{4y} = \frac{y^2}{y} = y.$$

24. Differentiating $P = ac_1 e^{at}/\left(1 + bc_1 e^{at}\right)$ we obtain

$$\frac{dP}{dt} = \frac{\left(1 + bc_1 e^{at}\right)a^2 c_1 e^{at} - ac_1 e^{at} \cdot abc_1 e^{at}}{(1 + bc_1 e^{at})^2}$$

$$= \frac{ac_1 e^{at}}{1 + bc_1 e^{at}} \cdot \frac{\left[a\left(1 + bc_1 e^{at}\right) - abc_1 e^{at}\right]}{1 + bc_1 e^{at}} = P(a - bP).$$

27. First write the differential equation in the form $y' = \dfrac{-x^2 - y^2}{x^2 - xy}$. Then $c_1(x + y)^2 = xe^{y/x}$ implies

$c_1 = \dfrac{xe^{y/x}}{(x + y)^2}$ and implicit differentiation gives $2c_1(x + y)(1 + y') = xe^{y/x}\dfrac{xy' - y}{x^2} + e^{y/x}$. Solving for y' we obtain

$$y' = \frac{e^{y/x} - \dfrac{y}{x}e^{y/x} - 2c_1(x + y)}{2c_1(x + y) - e^{y/x}} = \frac{1 - \dfrac{y}{x} - \dfrac{2x}{x + y}}{\dfrac{2x}{x + y} - 1} = \frac{-x^2 - y^2}{x^2 - xy}.$$

30. From $y = e^{2x} + xe^{2x}$ we obtain $\dfrac{dy}{dx} = 3e^{2x} + 2xe^{3x}$ and $\dfrac{d^2y}{dx^2} = 8e^{2x} + 4xe^{2x}$ so that $\dfrac{d^2y}{dx^2} - 4\dfrac{dy}{dx} + 4y = 0$.

33. From $y = \ln|x + c_1| + c_2$ we obtain $y' = \dfrac{1}{x + c_1}$ and $y'' = \dfrac{-1}{(x + c_1)^2}$, so that $y'' + (y')^2 = 0$.

36. From $y = x\cos(\ln x)$ we obtain $y' = -\sin(\ln x) + \cos(\ln x)$ and $y'' = \dfrac{-1}{x}\cos(\ln x) - \dfrac{1}{x}\sin(\ln x)$, so that $x^2 y'' - xy' + 2y = 0$.

39. From $y = x^2 e^x$ we obtain $y' = x^2 e^x + 2xe^x$, $y'' = x^2 e^x + 4xe^x 2e^x$, and $y''' = x^2 e^x + 6xe^x + 6e^x$, so that $y''' - 3y'' + 3y' - y = 0$.

42. From $y = \begin{cases} 0, & x < 0 \\ x^3, & x \ge 0 \end{cases}$ we obtain $y' = \begin{cases} 0, & x < 0 \\ 3x^2, & x \ge 0 \end{cases}$ so that $(y')^2 = \begin{cases} 0, & x < 0 \\ 9x^4, & x \ge 0. \end{cases}$

45. By inspection, $y = -1$ is a singular solution. Note that this is the "solution" obtained by computing the limit as c approaches infinity of the one-parameter family of solutions.

48. From $y = e^{mx}$ we obtain $y' = me^{mx}$ and $y'' = m^2 e^{mx}$. Then $y'' + 10y' + 25y = 0$ implies

$$m^2 e^{mx} + 10me^{mx} + 25e^{mx} = (m + 5)^2 e^{mx} = 0.$$

Since $e^{mx} > 0$ for all x, $m = -5$. Thus, $y = e^{-5x}$ is a solution.

51. It is easily shown that $y_1 = x^2$ and $y_2 = x^3$ are solutions. If $y_3 = c_1 y_1 + c_2 y_2 = c_1 x^2 + c_2 x^3$ then $y_3' = 2c_1 x + 3c_2 x^2$ and $y_3'' = 2c_1 + 6c_2 x$ so that $x^2 y_3'' - 4xy_3' + 6y_3 = 0$. Hence $c_1 y_1$, $c_2 y_2$, and $y_1 + y_2$ are solutions.

Exercises 1.2

3. (a) From $g = k/R^2$ we find $k = gR^2$.

(b) Using $a = \dfrac{d^2 r}{dt^2}$ and part (a) we obtain $\dfrac{d^2 r}{dt^2} = a = \dfrac{k}{r^2} = \dfrac{gR^2}{r^2}$ or $\dfrac{d^2 r}{dt^2} - \dfrac{gR^2}{r^2} = 0$.

(c) Part (b) becomes $\dfrac{dv}{dr}\dfrac{dr}{dt} - \dfrac{gR^2}{r^2} = 0$ or $v\dfrac{dv}{dr} - \dfrac{gR^2}{r^2} = 0$.

6. By Kirchoff's second law we obtain $R\dfrac{dq}{dt} + \dfrac{1}{C}q = E(t)$.

9. The differential equation is $\dfrac{dh}{dt} = -\dfrac{0.6 A_0}{A_w}\sqrt{2gh}$. We have $A_0 = \pi\left(\dfrac{1}{12}\right)^2 = \dfrac{\pi}{144}$ and $g = 32$. To find A_w we solve $x^2 + (5 - h)^2 = 25$ where x represents the radius of the circular area of the surface of the water whose depth is h. From $x = \sqrt{10h - h^2}$ we obtain $A_w = \pi(10h - h^2)$. Thus

$$\frac{dh}{dt} = -\frac{0.6\pi/144}{\pi(10h - h^2)}\sqrt{64h} = -\frac{1}{30h(10 - h)}\sqrt{h} = -\frac{1}{30\sqrt{h}\,(10 - h)}.$$

12. Equating Newton's law with the net forces in the x- and y-directions gives $m\dfrac{d^2x}{dt^2} = 0$ and $m\dfrac{d^2y}{dt^2} = -mg$, respectively.

15. To better understand the problem extend the line L down to the x-axis. Then we see from the figure that $\phi = 2\theta$, $\tan\phi = \dfrac{x}{y}$, and $\dfrac{dy}{dx} = \tan\left(\dfrac{\pi}{2} - \theta\right) = \cot\theta$. Now

$$\tan\phi = \tan 2\theta = \frac{2\tan\theta}{1-\tan^2\theta} = \frac{x}{y}, \text{ so } \frac{x}{y} = \frac{2(dx/dy)}{1-(dx/dy)^2} \text{ and } x\left(\frac{dx}{dy}\right)^2 + 2y\left(\frac{dx}{dy}\right) = x.$$

18. Substituting into the differential equation we obtain $-(m_0 - at)g = (m_0 - at)\dfrac{dv}{dt} + b(-a)$ or $(m_0 - at)\dfrac{dv}{dt} = ab - m_0 g + agt$.

21. The differential equation is $\dfrac{dA}{dt} = k(M - A)$.

Chapter 1 Review Exercises

3. Second-order; partial.

6. From $y = c_1 \cos(\ln x) + c_2 \sin(\ln x)$ we obtain $y' = \dfrac{1}{x}[c_2 \cos(\ln x) - c_1 \sin(\ln x)]$ and

$$y'' = \frac{-1}{x^2}[c_1 \cos(\ln x) + c_2 \sin(\ln x) + c_2 \cos(\ln x) - c_1 \sin(\ln x)]$$

so that $x^2 y'' + xy' + y = 0$.

9. $y = x^2$

12. $y = 2$

15. $y = \sin x$, $y = \cos x$, $y = 0$

18. If $|x| < 2$ and $|y| > 2$, then $(dy/dx)^2 < 0$ and the differential equation has no real solutions. This is also true for $|x| > 2$ and $|y| < 2$.

2 First-Order Differential Equations

―――――― **Exercises 2.1** ――――――――――――――――

3. For $f(x,y) = \dfrac{y}{x}$ we have $\dfrac{\partial f}{\partial y} = \dfrac{1}{x}$. Thus the differential equation will have a unique solution in any region where $x \neq 0$.

6. For $f(x,y) = \dfrac{x^2}{1+y^3}$ we have $\dfrac{\partial f}{\partial y} = \dfrac{-3x^2y^2}{(1+y^3)^2}$. Thus the differential equation will have a unique solution in any region where $y \neq -1$.

9. For $f(x,y) = x^3 \cos y$ we have $\dfrac{\partial f}{\partial y} = -x^2 \sin y$. Thus the differential equation will have a unique solution in the entire plane.

12. Two solutions are $y = 0$ and $y = x^2$. (Also, any constant multiple of x^2 is a solution.)

15. For $y = cx$ we have $y' = c$, from which we see that $y = cx$ is a solution of $xy' = y$ for all values of c. All of these solutions satisfy the initial condition $y(0) = 0$. The piecewise defined function is not a solution since it is not differentiable at $x = 0$.

18. We identify $f(x,y) = \sqrt{y^2 - 9}$ and $\partial f/\partial y = y^2/\sqrt{y^2 - 9}$. We then note that $f(x,y)$ is discontinuous for $|y| < 3$ and that $\partial f/\partial y$ is discontinuous for $|y| < 3$. Applying Theorem 2.1 we see that the differential equation is not guaranteed to have a unique solution at $(5,3)$.

―――――― **Exercises 2.2** ――――――――――――――――

In many of the following problems we will encounter an expression of the form $\ln |g(y)| = f(x) + c$. To solve for $g(y)$ we exponentiate both sides of the equation. This yields $|g(y)| = e^{f(x)+c} = e^c e^{f(x)}$ which implies $g(y) = \pm e^c e^{f(x)}$. Letting $c_1 = \pm e^c$ we obtain $g(y) = c_1 e^{f(x)}$.

3. From $dy = -e^{-3x} \, dx$ we obtain

$$y = \frac{1}{3} e^{-3x} + c.$$

6. From $dy = 2xe^{-x} dx$ we obtain

$$y = -2xe^{-x} - 2e^{-x} + c.$$

9. From $\dfrac{1}{y^3} \, dy = \dfrac{1}{x^2} \, dx$ we obtain

$$y^{-2} = \frac{2}{x} + c.$$

4

12. From $\left(\dfrac{1}{y} + 2y\right) dy = \sin x \, dx$ we obtain

$$\ln|y| + y^2 = -\cos x + c.$$

15. From $\dfrac{y}{2+y^2} dy = \dfrac{x}{4+x^2} dx$ we obtain

$$\ln|2+y^2| = \ln|4+x^2| + c \quad \text{or} \quad 2+y^2 = c_1\left(4+x^2\right).$$

18. From $\dfrac{y^2}{y+1} dy = \dfrac{1}{x^2} dx$ we obtain

$$\frac{1}{2}y^2 - y + \ln|y+1| = -\frac{1}{x} + c \quad \text{or} \quad \frac{1}{2}y^2 - y + \ln|y+1| = -\frac{1}{x} + c_1.$$

21. From $\dfrac{1}{S} dS = k \, dr$ we obtain $S = ce^{kr}$.

24. From $\dfrac{1}{N} dN = \left(te^{t+2} - 1\right) dt$ we obtain

$$\ln|N| = te^{t+2} - e^{t+2} - t + c.$$

27. From $\dfrac{e^{2y} - y}{e^y} dy = -\dfrac{\sin 2x}{\cos x} dx = -\dfrac{2\sin x \cos x}{\cos x} dx$ or $\left(e^y - ye^{-y}\right) dy = -2\sin x \, dx$ we obtain

$$e^y + ye^{-y} + e^{-y} = 2\cos x + c.$$

30. From $\dfrac{y}{\left(1+y^2\right)^{1/2}} dy = \dfrac{x}{\left(1+x^2\right)^{1/2}} dx$ we obtain

$$\left(1+y^2\right)^{1/2} = \left(1+x^2\right)^{1/2} + c.$$

33. From $\dfrac{y-2}{y+3} dy = \dfrac{x-1}{x+4} dx$ or $\left(1 - \dfrac{5}{y-3}\right) dy = \left(1 - \dfrac{5}{x+4}\right) dx$ we obtain

$$y - 5\ln|y-3| = x - 5\ln|x+4| + c \quad \text{or} \quad \left(\frac{x+4}{y-3}\right)^5 = c_1 e^{x-y}.$$

36. From $\sec y\dfrac{dy}{dx} + \sin x \cos y - \cos x \sin y = \sin x \cos y + \cos x \sin y$ we find $\sec y \, dy = 2\sin y \cos x \, dx$ or

$$\frac{1}{2\sin y \cos y} dy = \csc 2y \, dy = \cos x \, dx. \text{ Then}$$

$$\frac{1}{2}\ln|\csc 2y - \cot 2y| = \sin x + c.$$

39. From $\dfrac{1}{y^2} dy = \dfrac{1}{e^x + e^{-x}} dx = \dfrac{e^x}{(e^x)^2 + 1} dx$ we obtain

$$-\frac{1}{y} = \tan^{-1} e^x + c.$$

5

42. From $\dfrac{1}{1+(2y)^2}\,dy = \dfrac{-x}{1+(x^2)^2}\,dx$ we obtain

$$\frac{1}{2}\tan^{-1}2y = -\frac{1}{2}\tan^{-1}x^2 + c \quad \text{or} \quad \tan^{-1}2y + \tan^{-1}x^2 = c_1.$$

Using $y(1) = 0$ we find $c_1 = \pi/4$. The solution of the initial-value problem is

$$\tan^{-1}2y + \tan^{-1}x^2 = \frac{\pi}{4}.$$

45. From $\dfrac{1}{x^2+1}\,dx = 4\,dy$ we obtain $\tan^{-1}x = 4y + c$. Using $x(\pi/4) = 1$ we find $c = -3\pi/4$. The solution of the initial-value problem is

$$\tan^{-1}x = 4y - \frac{3\pi}{4} \quad \text{or} \quad x = \tan\left(4y - \frac{3\pi}{4}\right).$$

48. From $\dfrac{1}{1-2y}\,dy = dx$ we obtain $-\dfrac{1}{2}\ln|1-2y| = x + c$ or $1 - 2y = c_1 e^{-2x}$. Using $y(0) = 5/2$ we find $c_1 = -4$. The solution of the initial-value problem is

$$1 - 2y = -4e^{-2x} \quad \text{or} \quad y = 2e^{-2x} + \frac{1}{2}.$$

51. By inspection a singular solution is $y = 1$.

54. Separating variables we obtain $\dfrac{dy}{(y-1)^2} = dx$. Then $-\dfrac{1}{y-1} = x + c$ and $y = \dfrac{x+c-1}{x+c}$. Setting $x = 0$ and $y = 1.01$ we obtain $c = -100$. The solution is

$$y = \frac{x-101}{x-100}.$$

57. Let $u = x + y + 1$ so that $du/dx = 1 + dy/dx$. Then $\dfrac{du}{dx} - 1 = u^2$ or $\dfrac{1}{1+u^2}\,du = dx$. Thus $\tan^{-1}u = x + c$ or $u = \tan(x + c)$, and

$$x + y + 1 = \tan(x + c) \quad \text{or} \quad y = \tan(x + c) - x - 1.$$

60. Let $u = x + y$ so that $du/dx = 1 + dy/dx$. Then $\dfrac{du}{dx} - 1 = \sin u$ or $\dfrac{1}{1+\sin u}\,du = dx$. Multiplying by $(1 - \sin u)/(1 - \sin u)$ we have $\dfrac{1 - \sin u}{\cos^2 u}\,du = dx$ or $\left(\sec^2 u - \tan u \sec u\right)du = dx$. Thus

$$\tan u - \sec u = x + c \quad \text{or} \quad \tan(x + y) - \sec(x + y) = x + c.$$

Exercises 2.3

3. Since $f(tx, ty) = \dfrac{(tx)^3(ty) - (tx)^2(ty)^2}{(tx + 8ty)^2} = t^2 f(x, y)$, the function is homogeneous of degree 2.

6. Since $f(tx, ty) = \sin \dfrac{x}{x+y} = f(x, y)$, the function is homogeneous of degree 0.

9. Since $f(tx, ty) = \left(\dfrac{1}{tx} + \dfrac{1}{ty}\right)^2 = \dfrac{1}{t^2} f(x, y)$, the function is homogeneous of degree -2.

12. Letting $y = ux$ we have

$$(x + ux)\,dx + x(u\,dx + x\,du) = 0$$

$$(1 + 2u)\,dx + x\,du = 0$$

$$\frac{dx}{x} + \frac{du}{1 + 2u} = 0$$

$$\ln|x| + \frac{1}{2}\ln|1 + 2u| = c$$

$$x^2\left(1 + 2\frac{y}{x}\right) = c_1$$

$$x^2 + 2xy = c_1.$$

15. Letting $y = ux$ we have

$$\left(u^2 x^2 + ux^2\right)dx - x^2(u\,dx + x\,du) = 0$$

$$u^2\,dx - x\,du = 0$$

$$\frac{dx}{x} - \frac{du}{u^2} = 0$$

$$\ln|x| + \frac{1}{u} = c$$

$$\ln|x| + \frac{x}{y} = c$$

$$y\ln|x| + x = cy.$$

18. Letting $y = ux$ we have

$$(x + 3ux)\,dx - (3x + ux)(u\,dx + x\,du) = 0$$

$$\left(u^2 - 1\right)dx + x(u + 3)\,du = 0$$

$$\frac{dx}{x} + \frac{u + 3}{(u - 1)(u + 1)}\,du = 0$$

$$\ln|x| + 2\ln|u - 1| - \ln|u + 1| = c$$

$$\frac{x(u - 1)^2}{u + 1} = c_1$$

$$x\left(\frac{y}{x} - 1\right)^2 = c_1\left(\frac{y}{x} + 1\right)$$

$$(y - x)^2 = c_1(y + x).$$

21. Letting $x = vy$ we have

$$2v^2 y^3(v\,dy + y\,dv) - \left(3v^3 y^3 + y^3\right)dy = 0$$

$$2v^2 y\,dv - \left(v^3 + 1\right)dy = 0$$

$$\frac{2v^2}{v^3 + 1}\,dv - \frac{dy}{y} = 0$$

$$\frac{2}{3}\ln\left|v^3 + 1\right| - \ln|y| = c$$

$$\left(v^3 + 1\right)^{2/3} = c_1 y$$

$$\left(\frac{x^3}{y^3} + 1\right)^2 = c_2 y^3$$

$$\left(x^3 + y^3\right)^2 = c_2 y^9.$$

24. Letting $y = ux$ we have

$$\left(u^3x^3 + x^3 + u^2x^3\right) dx - u^2x^3\left(u\,dx + x\,du\right) = 0$$

$$\left(1 + u^2\right) dx - u^2x\,du = 0$$

$$\frac{dx}{x} - \frac{u^2}{u^2 + 1}\,du = 0$$

$$\ln|x| - u + \tan^{-1} u = c$$

$$\ln|x| - \frac{y}{x} + \tan^{-1}\frac{y}{x} = c.$$

27. Letting $y = ux$ we have

$$(ux + x\cot u)\,dx - x(u\,dx + x\,du) = 0$$

$$\cot u\,dx - x\,du = 0$$

$$\frac{dx}{x} - \tan u\,du = 0$$

$$\ln|x| + \ln|\cos u| = c$$

$$x\cos\frac{y}{x} = c.$$

30. Letting $y = ux$ we have

$$\left(x^2 + ux^2 + 3u^2x^2\right) dx - \left(x^2 + 2ux^2\right)\left(u\,dx + x\,du\right) = 0$$

$$\left(1 + u^2\right) dx - x(1 + 2u)\,du = 0$$

$$\frac{dx}{x} - \frac{1 + 2u}{1 + u^2}\,du = 0$$

$$\ln|x| - \tan^{-1} u - \ln\left(1 + u^2\right) = c$$

$$\frac{x}{1 + u^2} = c_1 e^{\tan^{-1} u}$$

$$x^3 = \left(y^2 + x^2\right) c_1 e^{\tan^{-1} y/x}.$$

33. Letting $y = ux$ we have

$$\left(3ux^2 + u^2x^2\right) dx - 2x^2(u\,dx + x\,du) = 0$$

$$\left(u^2 + u\right) dx - 2x\,du = 0$$

$$\frac{dx}{x} - \frac{2\,du}{u(u+1)} = 0$$

$$\ln|x| - 2\ln|u| + 2\ln|u+1| = c$$

$$\frac{x(u+1)^2}{u^2} = c_1$$

$$x\left(\frac{y}{x} + 1\right)^2 = c_1\left(\frac{y}{x}\right)^2$$

$$x(y+x)^2 = c_1 y^2.$$

Using $y(1) = -2$ we find $c_1 = 1/4$. The solution of the initial-value problem is $4x(y+x)^2 = y^2$.

36. Letting $x = vy$ we have

$$y(v\,dy + y\,dv) + (y\cos v - vy)\,dy = 0$$

$$y\,dv + \cos v\,dy = 0$$

$$\sec v\,dv + \frac{dy}{y} = 0$$

$$\ln|\sec v + \tan v| + \ln|y| = c$$

$$y\left(\sec\frac{x}{y} + \tan\frac{x}{y}\right) = c_1.$$

Using $y(0) = 2$ we find $c_1 = 2$. The solution of the initial-value problem is $y\left(\sec\dfrac{x}{y} + \tan\dfrac{x}{y}\right) = 2$.

39. Letting $y = ux$ we have

$$\left(x - ux - u^{3/2}x\right) dx + \left(x + \sqrt{u}\,x\right)(u\,dx + x\,du) = 0$$

$$dx + x\left(1 + \sqrt{u}\right) du = 0$$

$$\frac{dx}{x} + \left(1 + \sqrt{u}\right) du = 0$$

$$\ln x + u + \frac{2}{3}u^{3/2} = c$$

$$3x^{3/2}\ln x + 3x^{1/2}y + 2y^{3/2} = c_1 x^{3/2}.$$

Using $y(1) = 1$ we find $c_1 = 5$. The solution of the initial-value problem is

$$3x^{3/2} \ln x + 3x^{1/2} y + 2y^{3/2} = 5x^{3/2}.$$

(Note: Since the solution involves $\sqrt{x}$, $x \geq 0$ and we do not need an absolute value sign in $\ln x$.)

42. Letting $y = ux$ we have

$$\left(\sqrt{x} + \sqrt{ux} \right)^2 dx - x(u\, dx + x\, du) = 0$$

$$\left(1 + 2\sqrt{u} \right) dx - x\, du = 0$$

$$\frac{dx}{x} - \frac{du}{1 + 2\sqrt{u}} = 0$$

$$\ln |x| = \int \frac{du}{1 + 2\sqrt{u}} \qquad \boxed{u = t^2, \ du = 2t\, dt}$$

$$= \int \frac{2t}{1 + 2t}\, dt = t - \frac{1}{2} \ln |1 + 2t| + c$$

$$= \sqrt{\frac{y}{x}} - \frac{1}{2} \ln \left| 1 + 2\sqrt{\frac{y}{x}} \right| + c$$

$$x^2 \left(1 + 2\sqrt{\frac{y}{x}} \right) = c_1 e^{2\sqrt{y/x}}$$

$$x^{3/2} \left(\sqrt{x} + 2\sqrt{y} \right) = c_1 e^{2\sqrt{y/x}}.$$

Using $y(1) = 0$ we find $c_1 = 1$. The solution of the initial-value problem is

$$x^{3/2} \left(\sqrt{x} + 2\sqrt{y} \right) = e^{2\sqrt{y/x}}.$$

45. From $x = vy$ we obtain $dx = v\, dy + y\, dv$ and the differential equation becomes

$$M(vy, y)(v\, dy + y\, dv) + N(vy, y)\, dy = 0.$$

Using $M(vy, y) = y^n M(v, 1)$ and $N(vy, y) = y^2 N(v, 1)$ and simplifying we have

$$y^n M(v, 1)(v\, dy + y\, dv) + y^n N(v, 1)\, dy = 0$$

$$[vM(v, 1) + N(v, 1)]\, dy + yM(v, 1)\, dv = 0$$

$$\frac{dy}{y} + \frac{M(v, 1)\, dv}{vM(v, 1) + N(v, 1)} = 0.$$

48. If we let $u = y/x$, then by homogeneity $f(x, y) = x^n f\left(1, \frac{y}{x} \right) = x^n f(1, u)$. Using the chain rule for

11

partial derivatives, we obtain

$$\frac{\partial f(x,y)}{\partial x} = x^n \frac{\partial f(1,u)}{\partial u} \frac{\partial u}{\partial x} + nx^{n-1} f(1,u) = x^n \frac{\partial f(1,u)}{\partial u} \left(-\frac{y}{x^2}\right) + nx^{n-1} f(1,u)$$

$$= -yx^{n-2} \frac{\partial f(1,u)}{\partial u} + nx^{n-1} f(1,u)$$

and

$$\frac{\partial f(x,y)}{\partial y} = x^n \frac{\partial f(1,u)}{\partial u} \frac{\partial u}{\partial y} = x^n \frac{\partial f(1,u)}{\partial u} \left(\frac{1}{x}\right) = x^{n-1} \frac{\partial f(1,u)}{\partial u}.$$

Then

$$x \frac{\partial f}{\partial x} + y \frac{\partial f}{\partial y} = -yx^{n-1} \frac{\partial f(1,u)}{\partial u} + nx^n f(1,u) + yx^{n-1} \frac{\partial f(1,u)}{\partial u}$$

$$= nx^n f(1,u) = nx^n f\left(1, \frac{y}{x}\right) = nf(x,y).$$

Exercises 2.4

3. Let $M = 5x + 4y$ and $N = 4x - 8y^3$ so that $M_y = 4 = N_x$. From $f_x = 5x + 4y$ we obtain $f = \frac{5}{2}x^2 + 4xy + h(y)$, $h'(y) = -8y^3$, and $h(y) = -2y^4$. The solution is

$$\frac{5}{2}x^2 + 4xy - 2y^4 = c.$$

6. Let $M = 4x^3 - 3y \sin 3x - y/x^2$ and $N = 2y - 1/x + \cos 3x$ so that $M_y = -3 \sin 3x - 1/x^2$ and $N_x = 1/x^2 - 3 \sin 3x$. The equation is not exact.

9. Let $M = y^3 - y^2 \sin x - x$ and $N = 3xy^2 + 2y \cos x$ so that $M_y = 3y^2 - 2y \sin x = N_x$. From $f_x = y^3 - y^2 \sin x - x$ we obtain $f = xy^3 + y^2 \cos x - \frac{1}{2}x^2 + h(y)$, $h'(y) = 0$, and $h(y) = 0$. The solution is

$$xy^3 + y^2 \cos x - \frac{1}{2}x^2 = c.$$

12. Let $M = 2x/y$ and $N = -x^2/y^2$ so that $M_y = -2x/y^2 = N_x$. From $f_x = 2x/y$ we obtain $f = \frac{x^2}{y} + h(y)$, $h'(y) = 0$, and $h(y) = 0$. The solution is $x^2 = cy$.

15. Let $M = 1 - 3/x + y$ and $N = 1 - 3/y + x$ so that $M_y = 1 = N_x$. From $f_x = 1 - 3/x + y$ we obtain $f = x - 3 \ln|x| + xy + h(y)$, $h'(y) = 1 - \frac{3}{y}$, and $h(y) = y - 3 \ln|y|$. The solution is

$$x + y + xy - 3 \ln|xy| = c.$$

18. Let $M = -2y$ and $N = 5y - 2x$ so that $M_y = -2 = N_x$. From $f_x = -2y$ we obtain $f = -2xy + h(y)$, $h'(y) = 5y$, and $h(y) = \frac{5}{2}y^2$. The solution is

$$-2xy + \frac{5}{2}y^2 = c.$$

21. Let $M = 4x^3 + 4xy$ and $N = 2x^2 + 2y - 1$ so that $M_y = 4x = N_x$. From $f_x = 4x^3 + 4xy$ we obtain $f = x^4 + 2x^2y + h(y)$, $h'(y) = 2y - 1$, and $h(y) = y^2 - y$. The solution is

$$x^4 + 2x^2y + y^2 - y = c.$$

24. Let $M = 1/x + 1/x^2 - y/\left(x^2 + y^2\right)$ and $N = ye^y + x/\left(x^2 + y^2\right)$ so that $M_y = \left(y^2 - x^2\right)/\left(x^2 + y^2\right)^2 = N_x$. From $f_x = 1/x + 1/x^2 - y/\left(x^2 + y^2\right)$ we obtain $f = \ln|x| - \frac{1}{x} - \arctan\left(\frac{x}{y}\right) + h(y)$, $h'(y) = ye^y$, and $h(y) = ye^y - e^y$. The solution is

$$\ln|x| - \frac{1}{x} - \arctan\left(\frac{x}{y}\right) + ye^y - e^y = c.$$

27. Let $M = 4y + 2x - 5$ and $N = 6y + 4x - 1$ so that $M_y = 4 = N_x$. From $f_x = 4y + 2x - 5$ we obtain $f = 4xy + x^2 - 5x + h(y)$, $h'(y) = 6y - 1$, and $h(y) = 3y^2 - y$. The general solution is $4xy + x^2 - 5x + 3y^2 - y = c$. If $y(-1) = 2$ then $c = 8$ and the solution of the initial-value problem is

$$4xy + x^2 - 5x + 3y^2 - y = 8.$$

30. Let $M = y^2 + y\sin x$ and $N = 2xy - \cos x - 1/\left(1 + y^2\right)$ so that $M_y = 2y + \sin x = N_x$. From $f_x = y^2 + y\sin x$ we obtain $f = xy^2 - y\cos x + h(y)$, $h'(y) = \dfrac{-1}{1 + y^2}$, and $h(y) = -\tan^{-1}y$. The general solution is $xy^2 - y\cos x - \tan^{-1}y = c$. If $y(0) = 1$ then $c = -1 - \pi/4$ and the solution of the initial-value problem is

$$xy^2 - y\cos x - \tan^{-1}y = -1 - \frac{\pi}{4}.$$

33. Equating $M_y = 4xy + e^x$ and $N_x = 4xy + ke^x$ we obtain $k = 1$.

36. Since $f_x = M(x,y) = y^{1/2}x^{-1/2} + x\left(x^2 + y\right)^{-1}$ we obtain $f = 2y^{1/2}x^{1/2} + \frac{1}{2}\ln\left|x^2 + y\right| + h(x)$ so that $f_y = y^{-1/2}x^{1/2} + \frac{1}{2}\left(x^2 + y\right)^{-1} + h'(x)$. Let

$$N(x,y) = y^{-1/2}x^{1/2} + \frac{1}{2}\left(x^2 + y\right)^{-1}.$$

39. Let $M = -x^2y^2\sin x + 2xy^2\cos x$ and $N = 2x^2y\cos x$ so that $M_y = -2x^2y\sin x + 4xy\cos x = N_x$. From $f_y = 2x^2y\cos x$ we obtain $f = x^2y^2\cos x + h(y)$, $h'(y) = 0$, and $h(y) = 0$. The solution of the differential equation is

$$x^2y^2\cos x = c.$$

42. Let $M = \left(x^2 + 2xy - y^2\right) / \left(x^2 + 2xy + y^2\right)$ and $N = \left(y^2 + 2xy - x^2\right) / \left(y^2 + 2xy + x^2\right)$ so that $M_y = -4xy/(x+y)^3 = N_x$. From $f_x = \left(x^2 + 2xy + y^2 - 2y^2\right) / (x+y)^2$ we obtain

$f = x + \dfrac{2y^2}{x+y} + h(y)$, $h'(y) = -1$, and $h(y) = -y$. The solution of the differential equation is

$$x^2 + y^2 = c(x+y).$$

Exercises 2.5 —————————

3. For $y' + 4y = \dfrac{4}{3}$ an integrating factor is $e^{\int 4\,dx} = e^{4x}$ so that

$$\frac{d}{dx}\left[e^{4x}y\right] = \frac{4}{3}e^{4x} \quad \text{and} \quad y = \frac{1}{3} + ce^{-4x}$$

for $-\infty < x < \infty$.

6. For $y' - y = e^x$ an integrating factor is $e^{-\int dx} = e^{-x}$ so that

$$\frac{d}{dx}\left[e^{-x}y\right] = 1 \quad \text{and} \quad y = xe^x + ce^x$$

for $-\infty < x < \infty$.

9. For $y' + \dfrac{1}{x}y = \dfrac{1}{x^2}$ an integrating factor is $e^{\int (1/x)\,dx} = x$ so that

$$\frac{d}{dx}\left[xy\right] = \frac{1}{x} \quad \text{and} \quad y = \frac{1}{x}\ln x + \frac{c}{x}$$

for $0 < x < \infty$.

12. For $\dfrac{dx}{dy} - x = y$ an integrating factor is $e^{-\int dy} = e^{-y}$ so that

$$\frac{d}{dy}\left[e^{-y}x\right] = ye^{-y} \quad \text{and} \quad x = -y - 1 + ce^y$$

for $-\infty < y < \infty$.

15. For $y' + \dfrac{e^x}{1 + e^x}y = 0$ an integrating factor is $e^{\int [e^x/(1+e^x)]\,dx} = 1 + e^x$ so that

$$\frac{d}{dx}\left[1 + e^x y\right] = 0 \quad \text{and} \quad y = \frac{c}{1 + e^x}$$

for $-\infty < x < \infty$.

18. For $y' + (\cot x)y = 2\cos x$ an integrating factor is $e^{\int \cot x\,dx} = \sin x$ so that

$$\frac{d}{dx}\left[(\sin x)\,y\right] = 2\sin x \cos x \quad \text{and} \quad y = \sin x + c\csc x$$

for $0 < x < \pi$.

21. For $y' + \left(1 + \dfrac{2}{x}\right)y = \dfrac{e^x}{x^2}$ an integrating factor is $e^{\int [1+(2/x)]dx} = x^2 e^x$ so that

$$\frac{d}{dx}\left[x^2 e^x y\right] = e^{2x} \quad \text{and} \quad y = \frac{1}{2}\frac{e^x}{x^2} + \frac{ce^{-x}}{x^2}$$

for $0 < x < \infty$.

24. For $y' + \dfrac{2\sin x}{(1 - \cos x)}y = \tan x(1 - \cos x)$ an integrating factor is $e^{\int [2\sin x/(1-\cos x)]dx} = (1 - \cos x)^2$ so that

$$\frac{d}{dx}\left[(1 - \cos x)^2 y\right] = \tan x - \sin x \quad \text{and} \quad y(1 - \cos x)^2 = \ln|\sec x| + \cos x + c$$

for $0 < x < \pi/2$.

27. For $y' + \left(3 + \dfrac{1}{x}\right)y = \dfrac{e^{-3x}}{x}$ an integrating factor is $e^{\int [3+(1/x)]dx} = xe^{3x}$ so that

$$\frac{d}{dx}\left[xe^{3x}y\right] = 1 \quad \text{and} \quad y = e^{-3x} + \frac{ce^{-3x}}{x}$$

for $0 < x < \infty$.

30. For $y' + \dfrac{2}{x}y = \dfrac{1}{x}\left(e^x + \ln x\right)$ an integrating factor is $e^{\int (2/x)dx} = x^2$ so that

$$\frac{d}{dx}\left[x^2 y\right] = xe^x + x\ln x \quad \text{and} \quad x^2 y = xe^x - e^x + \frac{x^2}{2}\ln x - \frac{1}{4}x^2 + c$$

for $0 < x < \infty$.

33. For $\dfrac{dx}{dy} + \left(2y + \dfrac{1}{y}\right)x = 2$ an integrating factor is $e^{\int [2y+(1/y)]dy} = ye^{y^2}$ so that

$$\frac{d}{dy}\left[ye^{y^2}x\right] = 2ye^{y^2} \quad \text{and} \quad x = \frac{1}{y} + \frac{1}{y}ce^{-y^2}$$

for $0 < y < \infty$.

36. For $\dfrac{dP}{dt} + (2t - 1)P = 4t - 2$ an integrating factor is $e^{\int (2t-1)\,dt} = e^{t^2 - t}$ so that

$$\frac{d}{dt}\left[Pe^{t^2-t}\right] = (4t - 2)e^{t^2-t} \quad \text{and} \quad P = 2 + ce^{t-t^2}$$

for $-\infty < t < \infty$.

39. For $y' + (\cosh x)y = 10\cosh x$ an integrating factor is $e^{\int \cosh x\,dx} = e^{\sinh x}$ so that

$$\frac{d}{dx}\left[e^{\sinh x}y\right] = 10(\cosh x)e^{\sinh x} \quad \text{and} \quad y = 10 + ce^{-\sinh x}$$

for $-\infty < x < \infty$.

42. For $y' - 2y = x\left(e^{3x} - e^{2x}\right)$ an integrating factor is $e^{-\int 2dx} = e^{-2x}$ so that

$$\frac{d}{dx}\left[e^{-2x}y\right] = xe^x - x \quad \text{and} \quad y = xe^{3x} - e^{3x} - \frac{1}{2}x^2e^{2x} + ce^{2x}$$

for $-\infty < x < \infty$. If $y(0) = 2$ then $c = 3$ and

$$y = xe^{3x} - e^{3x} - \frac{1}{2}x^2e^{2x} + 3e^{2x}.$$

45. For $y' + (\tan x)y = \cos^2 x$ an integrating factor is $e^{\int \tan x \, dx} = \sec x$ so that

$$\frac{d}{dx}\left[(\sec x)\,y\right] = \cos x \quad \text{and} \quad y = \sin x \cos x + c\cos x$$

for $-\pi/2 < x < \pi/2$. If $y(0) = -1$ then $c = -1$ and

$$y = \sin x \cos x - \cos x.$$

48. For $y' + \left(1 + \dfrac{2}{x}\right)y = \dfrac{2}{x}e^{-x}$ an integrating factor is $e^{\int(1+2/x)dx} = x^2e^x$ so that

$$\frac{d}{dx}\left[x^2e^x y\right] = 2x \quad \text{and} \quad y = e^{-x} + \frac{c}{x^2}e^{-x}$$

for $0 < x < \infty$. If $y(1) = 0$ then $c = -1$ and

$$y = e^{-x} - \frac{1}{x^2}e^{-x}.$$

51. For $y' + \dfrac{2}{x(x-2)}y = 0$ an integrating factor is $e^{\int [2/x(x-2)]dx} = \dfrac{x-2}{x}$ so that

$$\frac{d}{dx}\left[\frac{x-2}{x}y\right] = 0 \quad \text{and} \quad (x-2)y = cx$$

for $2 < x < \infty$. If $y(3) = 6$ then $c = 2$ and

$$y = \frac{2x}{x-2}.$$

54. For $y' + \left(\sec^2 x\right)y = \sec^2 x$ an integrating factor is $e^{\int(\sec^2 x)dx} = e^{\tan x}$ so that

$$\frac{d}{dx}\left[e^{\tan x}y\right] = \sec^2 x\, e^{\tan x} \quad \text{and} \quad y = 1 + ce^{-\tan x}$$

for $-\pi/2 < x < \pi/2$. If $y(0) = -3$ then $c = -4$ and

$$y = 1 - 4e^{-\tan x}.$$

57. For $y' + 2xy = f(x)$ an integrating factor is e^{x^2} so that

$$ye^{x^2} = \begin{cases} \frac{1}{2}e^{x^2} + c_1, & 0 \le x < 1; \\ c_2, & x \ge 1. \end{cases}$$

If $y(0) = 2$ then $c_1 = 3/2$ and for continuity we must have $c_2 = \dfrac{1}{2}e + \dfrac{3}{2}$ so that

$$y = \begin{cases} \frac{1}{2} + \frac{3}{2}e^{-x^2}, & 0 \le x < 1; \\ \left(\frac{1}{2}e + \frac{3}{2}\right)e^{-x^2}, & x \ge 1. \end{cases}$$

Exercises 2.6

3. From $y' + y = xy^4$ and $w = y^{-3}$ we obtain $\dfrac{dw}{dx} - 3w = -3x$. An integrating factor is e^{-3x} so that

$$e^{-3x} = xe^{-3x} + \frac{1}{3}e^{-3x} + c \quad \text{or} \quad y^{-3} = x + \frac{1}{3} + ce^{3x}.$$

6. From $y' + \dfrac{2}{3\left(1 + x^2\right)}y = \dfrac{2x}{3\left(1 + x^2\right)}y^4$ and $w = y^{-3}$ we obtain $\dfrac{dw}{dx} - \dfrac{2x}{1 + x^2}w = \dfrac{-2x}{1 + x^2}$. An integrating factor is $\dfrac{1}{1 + x^2}$ so that

$$\frac{w}{1 + x^2} = \frac{1}{1 + x^2} + c \quad \text{or} \quad y^{-3} = 1 + c\left(1 + x^2\right).$$

9. From $\dfrac{dx}{dy} - yx = y^3 x^2$ and $w = x^{-1}$ we obtain $\dfrac{dw}{dy} + yx = -y^3$. An integrating factor is $e^{y^2/2}$ so that

$$e^{y^2/2}w = -ye^{y^2/2} + 2e^{y^2/2} + c \quad \text{or} \quad x^{-1} = 2 - y^2 + ce^{-y^2/2}.$$

If $y(1) = 0$ then $c = -1$ and

$$x^{-1} = 2 - y^2 - e^{-y^2/2}.$$

12. Identify $P(x) = 1 - x$, $Q(x) = -1$, and $R(x) = x$. Then $\dfrac{dw}{dx} + (-1 + 2x)w = -x$. An integrating factor is $e^{x^2 - x}$ so that

$$e^{x^2 - x}w = -\int xe^{x^2 - x}dx + c \quad \text{or} \quad u = \frac{-e^{x^2 - x}}{\int xe^{x^2 - x}dx + c}.$$

Thus, $y = 1 + u$.

15. Identify $P(x) = e^{2x}$, $Q(x) = 1 + 2e^x$, and $R(x) = 1$. Then $\dfrac{dw}{dx} + (1 + 2e^x - 2e^x)w = -1$. An integrating factor is e^x so that

$$e^x w = -e^x + c \quad \text{or} \quad u = \frac{1}{ce^{-x} - 1}.$$

Thus, $y = -e^x + u$.

18. Identify $P(x) = 9$, $Q(x) = 6$, $R(x) = 1$, and $y_1 = -3$. An integrating factor for $\dfrac{dw}{dx} + (6-6)w = -1$ is 1 so that

$$w = -x + c \quad \text{or} \quad u = \frac{1}{-x+c}.$$

Thus, $y = -3 + u$.

21. Let $y = xy' + f(y')$ where $f(t) = -t^3$. A family of solutions is $y = cx - c^3$. The singular solution is given by

$$x = 3t^2 \quad \text{and} \quad y = 2t^3 \quad \text{or} \quad 27y^2 = 4x^3.$$

24. Let $y = xy' + f(y')$ where $f(t) = \ln t$. A family of solutions is $y = cx + \ln c$. The singular solution is given by

$$x = -\frac{1}{t} \quad \text{and} \quad y = \ln t - 1 \quad \text{or} \quad y = \ln\left(-\frac{1}{x}\right) - 1.$$

27. If $y' + y^2 - Q(x)y - P(x) = 0$ and $y = \dfrac{w'}{w}$ then $\dfrac{dy}{dx} = \dfrac{ww'' - w'w'}{w^2}$ and $w'' - Q(x)w' - P(x)w = 0$.

30. From $x = -f'(t)$ and $y = f(t) - tf'(t)$ we obtain

$$\frac{dy}{dx} = \frac{dy/dt}{dx/dt} = \frac{-tf''(t)}{-f''(t)} = t$$

for $f''(t) \neq 0$. Substituting into $y = xy' + f(y')$ we find $f(t) - tf'(t) = xt + f(t)$. Since $x = -f'(t)$, this becomes $f(t) - tf'(t) = -tf'(t) + f(t)$, which is an identity. Thus, the parametric equations form a solution of $y = xy' + f(y')$.

Exercises 2.7

3. Let $u = ye^x$. Then $y = ue^{-x}$ and $dy = -ue^{-x}\,dx + e^{-x}\,du$, and the equation becomes

$$ue^{-x}dx + (1+u)(-ue^{-x}dx + e^{-x}du) = 0 \quad \text{or} \quad (1+u)\,du = u^2dx.$$

Separating variables and integrating we find

$$-\frac{1}{u} + \ln|u| = x + c \implies -\frac{1}{ye^x} + \ln|y| + x = x + c \implies y\ln|y| = e^{-x} + cy.$$

6. Let $u = x + y$ so that $\dfrac{du}{dx} = 1 + \dfrac{dy}{dx}$. The equation becomes

$$\left(\frac{du}{dx} - 1\right) + u + 1 = u^2 e^{3x} \quad \text{or} \quad \frac{du}{dx} + u = u^2 e^{3x}.$$

This is a Bernoulli equation and we use the substitution $w = u^{-1}$ to obtain $\dfrac{dw}{dx} - w = e^{-3x}$. An integrating factor is e^{-x}, so

$$\frac{d}{dx}[e^{-x}w] = e^{-2x} \implies w = -\frac{1}{2}e^{3x} + ce^x \implies u = \frac{1}{-\frac{1}{2}e^{3x} + ce^x} \implies y = \frac{2}{-e^{3x} + c_1 e^x} - x.$$

9. Let $u = \ln(\tan y)$ so that $\dfrac{du}{dx} = \dfrac{\sec^2 y}{\tan y}\dfrac{dy}{dx} = 2\csc 2y\dfrac{dy}{dx}$. The equation becomes $x\dfrac{du}{dx} = 2x - u$ or $\dfrac{du}{dx} + \dfrac{1}{x}u = 2$. An integrating factor is x, so

$$\frac{d}{dx}[xu] = 2x \implies u = x + \frac{c}{x} \implies \ln(\tan y) = x + \frac{c}{x}.$$

12. Let $u = e^y$ so that $u' = e^y y'$. The equation becomes $xu' - 2u = x^2$ or $u' - \dfrac{2}{x}u = x$. An integrating factor is x^{-2}, so

$$\frac{d}{dx}\left[x^{-2}u\right] = \frac{1}{x} \implies u = x^2 \ln|x| + cx^2 \implies e^y = x^2 \ln|x| + cx^2.$$

15. Let $u = y^2 \ln x$ so that $\dfrac{du}{dy} = \dfrac{y^2}{x}\dfrac{dx}{dy} + 2y\ln x$ or $\dfrac{x}{y}\dfrac{du}{dy} = y\dfrac{dy}{dx} + 2x\ln x$. The equation becomes $\dfrac{x}{y}\dfrac{du}{dy} = xe^y$ or $\dfrac{1}{y}\dfrac{du}{dy} = e^y$. Separating variables we have

$$du = ye^y \implies u = ye^y - e^y + c \implies y^2 \ln x = ye^y - e^y + c.$$

18. Let $u = y'$ so that $u' = y''$. The equation becomes $u' - \dfrac{1}{x}u = u^2$, which is Bernoulli. Using the substitution $w = u^{-1}$ we obtain $\dfrac{dw}{dx} + \dfrac{1}{x}w = -1$. An integrating factor is x, so

$$\frac{d}{dx}[xw] = -x \implies w = -\frac{1}{2}x + \frac{1}{x}c \implies \frac{1}{u} = \frac{c_1 - x^2}{2x} \implies u = \frac{2x}{c_1 - x^2} \implies y = -\ln\left|c_1 - x^2\right| + c_2.$$

21. Let $u = y'$ so that $u' = y''$. The equation becomes $u = xu' + (u')^3 + 1$. This is a Clairaut equation with $f(t) = 1 + t^3$. A family of solutions is

$$u = c_1 x + \left(1 + c_1^3\right) \quad \text{and} \quad y = \frac{1}{2}c_1 x^2 + \left(1 + c_1^3\right)x + c_2.$$

A singular solution is given by $x = -3t^2$ and $u = 1 + t^3 - t\left(-3t^2\right) = 1 + 4t^3$. Eliminating the parameter we obtain

$$u = 1 + 4\left(-\frac{x}{3}\right)^{3/2} \quad \text{and} \quad y = x - \frac{24}{5}\left(-\frac{x}{3}\right)^{5/2}.$$

24. Let $u = y'$ so that $u' = y''$. The equation becomes $u' + u \tan x = 0$. Separating variables we obtain

$$\frac{du}{u} = -(\tan x)\,dx \implies \ln|u| = \ln|\cos x| + c \implies u = c_1 \cos x \implies y = c_1 \sin x + c_2.$$

27. We need to solve $\left[1 + (y')^2\right]^{3/2} = y''$. Let $u = y'$ so that $u' = y''$. The equation becomes $\left(1 + u^2\right)^{3/2} = u'$ or $\left(1 + u^2\right)^{3/2} = \dfrac{du}{dx}$. Separating variables and using the substitution $u = \tan\theta$ we have

$$\frac{du}{(1 + u^2)^{3/2}} = dx \implies \int \frac{\sec^2\theta}{\left(1 + \tan^2\theta\right)^{3/2}}\,d\theta = x \implies \int \frac{\sec^2\theta}{\sec^3\theta}\,d\theta = x$$

$$\implies \int \cos\theta\,d\theta = x \implies \sin\theta = x \implies \frac{u}{\sqrt{1 + u^2}} = x$$

$$\implies \frac{y'}{\sqrt{1 + (y')^2}} = x \implies (y')^2 = x^2\left[1 + (y')^2\right] = \frac{x^2}{1 - x^2}$$

$$\implies y' = \frac{x}{\sqrt{1 - x^2}} \quad \text{(for } x > 0\text{)} \implies y = -\sqrt{1 - x^2}\,.$$

Exercises 2.8

3. Identify $x_0 = 0$, $y_0 = 1$, and $f(t, y_{n-1}(t)) = 2t y_{n-1}(t)$. Picard's formula is

$$y_n(x) = 1 + 2\int_0^x t y_{n-1}(t)\,dt$$

for $n = 1, 2, 3, \ldots$. Iterating we find

$$y_1(x) = 1 + x^2 \qquad\qquad y_3(x) = 1 + x^2 + \frac{1}{2}x^4 + \frac{1}{6}x^6$$

$$y_2(x) = 1 + x^2 + \frac{1}{2}x^4 \qquad\qquad y_4(x) = 1 + x^2 + \frac{1}{2}x^4 + \frac{1}{6}x^6 + \frac{1}{24}x^8.$$

As $n \to \infty$, $y_n(x) \to e^{x^2}$.

6. Identify $x_0 = 0$, $y_0 = 1$, and $f(t, y_{n-1}(t)) = 2e^t - y_{n-1}(t)$. Picard's formula is

$$y_n(x) = 2e^x - 1 - \int_0^x y_{n-1}(t)\,dt$$

for $n = 1, 2, 3, \ldots$. Iterating we find

$$y_1(x) = 2e^x - 1 - x \qquad\qquad y_3(x) = 2e^x - 1 - x - \frac{1}{2}x^2 - \frac{1}{6}x^3$$

$$y_2(x) = 1 + x + \frac{1}{2}x^2 \qquad\qquad y_4(x) = 1 + x + \frac{1}{2}x^2 + \frac{1}{6}x^3 + \frac{1}{24}x^4.$$

As $n \to \infty$, $y_n(x) \to e^x$.

Chapter 2 Review Exercises

3. False; since $y = 0$ is a solution.

6. Separating variables we obtain

$$\cos^2 x \, dx = \frac{y}{y^2+1} \, dy \implies \frac{1}{2}x + \frac{1}{4}\sin 2x = \frac{1}{2}\ln\left(y^2+1\right) + c \implies 2x + \sin 2x = 2\ln\left(y^2+1\right) + c.$$

9. The equation is homogeneous, so let $y = ux$. Then $dy = u \, dx + x \, du$ and the differential equation becomes $ux^2(u \, dx + x \, du) = \left(3u^2x^2 + x^2\right) dx$ or $ux \, du = \left(2u^2 + 1\right) dx$. Separating variables we obtain

$$\frac{u}{2u^2+1} \, du = \frac{dx}{x} \implies \frac{1}{4}\ln\left(2u^2+1\right) = \ln x + c \implies 2u^2 + 1 = c_1 x^4$$

$$\implies 2\frac{y^2}{x^2} + 1 = c_1 x^4 \implies 2y^2 + x^2 = c_1 x^6.$$

If $y(-1) = 2$ then $c_1 = 9$ and the solution of the initial-value problem is $2y^2 + x^2 = 9x^6$.

12. Let $u = xy$ so that $du = x \, dy + y \, dx$. The differential equation becomes

$$du - y \, dx + \left(u + y - x^2 - 2x\right) dx = 0 \quad \text{or} \quad \frac{du}{dx} + u = x^2 + 2x.$$

An integrating factor is e^x, so

$$\frac{d}{dx}[e^x u] = \left(x^2 + 2x\right) e^x \implies e^x u = x^2 e^x + c \implies y = x + \frac{c}{x}e^{-x}.$$

15. The differential equation is Bernoulli. Using $w = y^{-1}$ we obtain $-xy^2 \dfrac{dw}{dx} + 4y = x^4 y^2$ or $\dfrac{dw}{dx} - \dfrac{4}{x}w = -x^3$. An integrating factor is x^{-4}, so

$$\frac{d}{dx}\left[x^{-4}w\right] = -\frac{1}{x} \implies x^{-4}w = -\ln x + c \implies w = -x^4 \ln x + cx^4 \implies y = \left(cx^4 - x^4 \ln x\right)^{-1}.$$

If $y(1) = 1$ then $c = 1$ and $y = \left(x^4 - x^4 \ln x\right)^{-1}$.

18. Let $u = y'$ so that $u' = y''$. The equation becomes $u' = x - u$ or $u' + u = x$. An integrating factor is e^x, so

$$\frac{d}{dx}[e^x u] = xe^x \implies e^x u = xe^x - e^x + c_1 \implies y' = x - 1 + c_1 e^{-x} \implies y = \frac{1}{2}x^2 - x - c_1 e^{-x} + c_2.$$

3 Applications of First-Order Differential Equations

—————— **Exercises 3.1** ——————

3. From $y = c_1 x^2$ we obtain $y' = \dfrac{2y}{x}$ so that the differential equation of the orthogonal family is $y' = -\dfrac{x}{2y}$. Then

$$2y\,dy = -x\,dx \quad \text{and} \quad 2y^2 + x^2 = c_2.$$

6. From $2x^2 + y^2 = c_1^2$ we obtain $y' = -\dfrac{2x}{y}$ so that the differential equation of the orthogonal family is $y' = \dfrac{y}{2x}$. Then

$$\frac{1}{y}\,dy = \frac{1}{2x}\,dx \quad \text{and} \quad y^2 = c_2 x.$$

9. From $y^2 = c_1 x^3$ we obtain $y' = \dfrac{3y}{2x}$ so that the differential equation of the orthogonal family is $y' = -\dfrac{2x}{3y}$. Then

$$3y\,dy = -2x\,dx \quad \text{and} \quad 3y^2 + 2x^2 = c_2.$$

12. From $y = \dfrac{1 + c_1 x}{1 - c_1 x}$ we obtain $y' = \dfrac{y^2 - 1}{2x}$ so that the differential equation of the orthogonal family is $y' = \dfrac{2x}{1 - y^2}$. Then

$$\left(1 - y^2\right) dy = 2x\,dx \quad \text{and} \quad 3y - 3x^2 - y^3 = c_2.$$

15. From $y^3 + 3x^2 y = c_1$ we obtain $y' = -\dfrac{2xy}{x^2 + y^2}$ so that the differential equation of the orthogonal family is $y' = \dfrac{x^2 + y^2}{2xy}$. This is a homogeneous differential equation. Let $y = ux$ so that $y' = u + xu'$. Then

$$\frac{2u}{1 - u^2}\,du = \frac{dx}{x} \implies -\ln\left|1 - u^2\right| = \ln|x| + c \implies x\left(1 - \frac{y^2}{x^2}\right) = c_1 \implies x^2 - y^2 = c_1 x.$$

18. From $y = \dfrac{1}{c_1 + x}$ we obtain $y' = -y^2$ so that the differential equation of the orthogonal family is $y' = \dfrac{1}{y^2}$. Then

$$y^2\,dy = dx \quad \text{and} \quad y^3 = 3x + c.$$

21. From $y = \dfrac{1}{\ln c_1 x}$ we obtain $y' = -\dfrac{y^2}{x}$ so that the differential equation of the orthogonal family is $y' = \dfrac{x}{y^2}$. Then

$$y^2 \, dy = x \, dx \quad \text{and} \quad 2y^3 = 3x^2 + c.$$

24. From $y = c_1 \sin x$ we obtain $y' = y \cot x$ so that the differential equation of the orthogonal family is $y' = -\dfrac{\tan x}{y}$. Then

$$y \, dy = -\tan x \, dx \quad \text{and} \quad y^2 = 2 \ln |\cos x| + c_2.$$

27. From $x + y = c_1 e^y$ we obtain $y' = \dfrac{1}{x + y - 1}$ so that the differential equation of the orthogonal family is $y' = 1 - x - y$. Then $y' + y = 1 - x$. An integrating factor is e^x, so

$$\frac{d}{dx}[e^x y] = e^x - xe^x \implies e^x y = 2e^x - xe^x + c \implies y = 2 - x + ce^{-x}.$$

If $y(0) = 5$ then $c = 3$ and $y = 2 - x + 3e^{-x}$.

30. From $r = c_1(1 + \cos\theta)$ we obtain $r \dfrac{d\theta}{dr} = -\dfrac{1 + \cos\theta}{\sin\theta}$ so that the differential equation of the orthogonal family is $r \dfrac{d\theta}{dr} = \dfrac{\sin\theta}{1 + \cos\theta}$. Then

$$\frac{1 + \cos\theta}{\sin\theta} \, d\theta = \frac{dr}{r} \implies \frac{\sin\theta}{1 - \cos\theta} \, d\theta = \frac{dr}{r} \implies \ln|1 - \cos\theta| = \ln|r| + c \implies r = c_1(1 - \cos\theta).$$

33. From $r = c_1 \sec\theta$ we obtain $r \dfrac{d\theta}{dr} = \cot\theta$ so that the differential equation of the orthogonal family is $r \dfrac{d\theta}{dr} = -\tan\theta$. Then

$$-\cot\theta = \frac{dr}{r} \implies -\ln|\sin\theta| = \ln|r| + c \implies r = c_1 \csc\theta.$$

36. Since the differential equation of the original family is $f(x, y) = \dfrac{y}{x}$, the differential equation of the isogonal family is $y' = \dfrac{y/x \pm 1}{1 \mp y/x} = \dfrac{y \pm x}{x \mp y}$. This is homogeneous so let $y = ux$. Then $y' = u + xu'$ and

$$xu' = \frac{\pm 1 \pm u^2}{1 \mp u} \implies \pm \frac{1 \mp u}{1 + u^2} \, du = \frac{dx}{x} \implies \pm \tan^{-1} u - \frac{1}{2} \ln\left(1 + u^2\right) = \ln|x| + c$$

$$\implies \pm 2 \tan^{-1} \frac{y}{x} - \ln\left(1 + \frac{y^2}{x^2}\right) = 2\ln|x| + c_1 \implies \pm 2 \tan \frac{y}{x} - \ln\left(x^2 + y^2\right) = c_1.$$

23

39. From $y^2 = c_1(2x + c_1)$ we obtain $c_1 = -x \pm \sqrt{x^2 + y^2}$ and

$$y' = -\frac{x}{y} + \sqrt{\left(\frac{x}{y}\right)^2 + 1} \quad \text{or} \quad y' = -\frac{x}{y} - \sqrt{\left(\frac{x}{y}\right)^2 + 1}.$$

Self–orthogonality follows from the fact that the product of these derivatives is -1.

42. We have $\psi_1 - \psi_2 = \dfrac{\pi}{2}$ so that $\tan \psi_1 = \tan\left(\psi_2 + \dfrac{\pi}{2}\right) = -\cot \psi_2 = -\dfrac{1}{\tan \psi_2}$.

Exercises 3.2

3. Let $P = P(t)$ be the population at time t. From $dP/dt = kt$ and $P(0) = P_0 = 500$ we obtain $P = 500e^{kt}$. Using $P(10) = 575$ we find $k = \frac{1}{10}\ln 1.15$. Then $P(30) = 500e^{3\ln 1.15} \approx 760$ years.

6. Let $N = N(t)$ be the amount at time t. From $dN/dt = kt$ and $N(0) = 100$ we obtain $N = 100e^{kt}$. Using $N(6) = 97$ we find $k = \frac{1}{6}\ln 0.97$. Then $N(24) = 100e^{(1/6)(\ln 0.97)24} = 100(0.97)^4 \approx 88.5$ mg.

9. Let $I = I(t)$ be the intensity, t the thickness, and $I(0) = I_0$. If $dI/dt = kI$ and $I(3) = .25I_0$ then $I = I_0 e^{kt}$, $k = \frac{1}{3}\ln .25$, and $I(15) = .00098I_0$.

12. Assume that $dT/dt = k(T - 5)$ so that $T = 5 + ce^{kt}$. If $T(1) = 55°$ and $T(5) = 30°$ then $k = -\frac{1}{4}\ln 2$ and $c = 59.4611$ so that $T(0) = 64.4611°$.

15. Assume $L\,di/dt + Ri = E(t)$, $L = .1$, $R = 50$, and $E(t) = 50$ so that $i = \frac{3}{5} + ce^{-500t}$. If $i(0) = 0$ then $c = -3/5$ and $\lim_{t \to \infty} i(t) = 3/5$.

18. Assume $R\,dq/dt + (1/c)q = E(t)$, $R = 1000$, $C = 5 \times 10^{-6}$, and $E(t) = 200$ so that $q = 1/1000 + ce^{-200t}$ and $i = -200ce^{-200t}$. If $i(0) = .4$ then $c = -1/500$, $q(.005) = .003$ coulombs, and $i(.005) = .1472$ amps. As $t \to \infty$ we have $q \to 1/1000$.

21. From $dA/dt = 4 - A/50$ we obtain $A = 200 + ce^{-t/50}$. If $A(0) = 30$ then $c = -170$ and $A = 200 - 170e^{-t/50}$.

24. From $\dfrac{dA}{dt} = 10 - \dfrac{10A}{500 - (10 - 5)t} = 10 - \dfrac{2A}{100 - t}$ we obtain $A = 1000 - 10t + c(100 - t)^2$. If $A(0) = 0$ then $c = -\dfrac{1}{10}$. The tank is empty in 100 minutes.

27. (a) From $m\,dv/dt = mg - kv$ we obtain $v = gm/k + ce^{-kt/m}$. If $v(0) = v_0$ then $c = v_0 - gm/k$ and the solution of the initial-value problem is

$$v = \frac{gm}{k} + \left(v_0 - \frac{gm}{k}\right)e^{-kt/m}.$$

(b) As $t \to \infty$ the limiting velocity is gm/k.

(c) From $ds/dt = v$ and $s(0) = s_0$ we obtain

$$s = \frac{gm}{k}t - \frac{m}{k}\left(v_0 - \frac{gm}{k}\right)e^{-kt/m} + s_0 + \frac{m}{k}\left(v_0 - \frac{gm}{k}\right).$$

30. From $V\,dC/dt = kA(C_s - C)$ and $C(0) = C_0$ we obtain $C = C_s + (C_0 - C_s)e^{-kAt/V}$.

33. From $r^2 d\theta = (L/m)\,dt$ we obtain $A = \dfrac{1}{2}\displaystyle\int_{\theta_1}^{\theta_2} r^2 d\theta = \dfrac{1}{2}\dfrac{L}{m}\displaystyle\int_a^b dt = \dfrac{1}{2}\dfrac{L}{m}(b - a).$

Exercises 3.3

3. From $\dfrac{dP}{dt} = P\left(10^{-1} - 10^{-7}P\right)$ and $P(0) = 5000$ we obtain $P = \dfrac{500}{.0005 + .0995e^{-.1t}}$ so that

$P \to 1{,}000{,}000$ as $t \to \infty$. If $P(t) = 500{,}000$ then $t = 52.9\,\text{months}$.

6. From Problem 5 we have $P = e^{a/b}e^{-ce^{-bt}}$ so that

$$\frac{dP}{dt} = bce^{a/b-bt}e^{-ce^{-bt}} \quad\text{and}\quad \frac{d^2P}{dt^2} = b^2 ce^{a/b-bt}e^{-ce^{-bt}}\left(ce^{-bt} - 1\right).$$

Setting $d^2P/dt^2 = 0$ and using $c = a/b - \ln P_0$ we obtain $t = (1/b)\ln(a/b - \ln P_0)$ and $P = e^{a/b-1}$.

9. If $\alpha \neq \beta$, $\dfrac{dX}{dt} = k(\alpha - X)(\beta - X)$, and $X(0) = 0$ then $\left(\dfrac{1/(\beta - \alpha)}{\alpha - X} + \dfrac{1/(\alpha - \beta)}{\beta - X}\right)dX = k\,dt$ so that

$X = \dfrac{\alpha\beta - \alpha\beta e^{(\alpha-\beta)kt}}{\beta - \alpha e^{(\alpha-\beta)kt}}$. If $\alpha = \beta$ then $\dfrac{1}{(\alpha - X)^2}\,dX = k\,dt$ and $X = \alpha - \dfrac{1}{kt + c}$.

12. From $\dfrac{d^2y}{dx^2} = \dfrac{w}{T_1}\sqrt{1 + \left(\dfrac{dy}{dx}\right)^2}$, $p = \dfrac{dy}{dx}$, and $y'(0) = 0$ we obtain $p + \sqrt{1 + p^2} = e^{wx/T_1}$ so that

$p = \sinh\dfrac{w}{T_1}x$. From $y(0) = 1$ it follows that $y = \dfrac{T_1}{w}\cosh\dfrac{w}{T_1}x + 1 - \dfrac{T_1}{w}$.

15. From $\dfrac{dh}{dt} = -\dfrac{\sqrt{h}}{25}$ and $h(0) = 20$ we obtain $h = \left(\sqrt{20} - \dfrac{t}{50}\right)^2$. If $h(t) = 0$ then $t = 50\sqrt{20}\,\text{seconds}$.

18. From $m\dfrac{dv}{dt} = mg - kv^2$ and $v(0) = v_0$ we obtain

$$\left[\frac{1/2g}{1 - \sqrt{k/mg}\,v} + \frac{1/2g}{1 + \sqrt{k/mg}\,v}\right]dv = dt$$

so that

$$\frac{v + \sqrt{mg/k}}{v - \sqrt{mg/k}} = \frac{v_0 + \sqrt{mg/k}}{v_0 - \sqrt{mg/k}}e^{2\sqrt{gk/m}\,t}.$$

Divide this equation by $e^{2\sqrt{gk/m}\,t}$ and multiply by $v - \sqrt{mg/k}$ to see that $v \to \sqrt{mg/k}$ as $t \to \infty$.

21. Using $\dfrac{dy}{dx} = \dfrac{dy}{dt} \Big/ \dfrac{dx}{dt}$ we obtain $\left(\dfrac{-\gamma + \delta y}{y}\right) dy = \left(\dfrac{\alpha - \beta x}{x}\right) dx$. Using $x \geq 0$ and $y \geq 0$ we have $-\gamma \ln y + \delta y = \alpha \ln x - \beta x + c$.

——— Chapter 3 Review Exercises ———

3. From $y - 2 = c_1(x - 1)^2$ we obtain $y' = \dfrac{2(y - 2)}{x - 1}$ so that the differential equation of the orthogonal family is $y' = \dfrac{1 - x}{2(y - 2)}$. The orthogonal trajectories are $(y - 2)^2 = x - \dfrac{1}{2}x^2 + c_2$.

6. Let $A = A(t)$ be the volume of CO_2 at time t. From $\dfrac{dA}{dt} = 1.2 - \dfrac{A}{4}$ and $A(0) = 16\,\text{ft}^3$ we obtain $A = 4.8 + 11.2e^{-t/4}$. Since $A(10) = 5.7\,\text{ft}^3$, the concentration is 0.017%. As $t \to \infty$ we have $A \to 4.8\,\text{ft}^3$ or 0.06%.

9. (a) The differential equation is

$$\frac{dT}{dt} = k[T - T_2 - B(T_1 - T)] = k[(1 + B)T - (BT_1 + T_2)].$$

Separating variables we obtain $\dfrac{dT}{(1 + B)T - (BT_1 + T_2)} = k\,dt$. Then

$$\frac{1}{1 + B} \ln|(1 + B)T - (BT_1 + T_2)| = kt + c \quad \text{and} \quad T(t) = \frac{BT_1 + T_2}{1 + B} + c_3 e^{k(1+B)t}.$$

Since $T(0) = T_1$ we must have $c_3 = \dfrac{T_1 - T_2}{1 + B}$ and so

$$T(t) = \frac{BT_1 + T_2}{1 + B} + \frac{T_1 - T_2}{1 + B}e^{k(1+B)t}.$$

(b) Since $k < 0$, $\displaystyle\lim_{t\to\infty} e^{k(1+B)t} = 0$ and $\displaystyle\lim_{t\to\infty} T(t) = \dfrac{BT_1 + T_2}{1 + B}$.

(c) Since $T_s = T_2 + B(T_1 - T)$, $\displaystyle\lim_{t\to\infty} T_s = T_2 + BT_1 - B\left(\dfrac{BT_1 + T_2}{1 + B}\right) = \dfrac{BT_1 + T_2}{1 + B}$.

4 Linear Differential Equations of Higher Order

Exercises 4.1

3. From $y = c_1 e^{4x} + c_2 e^{-x}$ we find $y' = 4c_1 e^{4x} - c_2 e^{-x}$. Then $y(0) = c_1 + c_2 = 1$, $y'(0) = 4c_1 - c_2 = 2$ so that $c_1 = 3/5$ and $c_2 = 2/5$. The solution is

$$y = \frac{3}{5} e^{4x} + \frac{2}{5} e^{-x}.$$

6. From $y = c_1 + c_2 x^2$ we find $y' = 2c_2 x$. Then $y(0) = c_1 = 0$, $y'(0) = 2c_2 \cdot 0 = 0$ and $y'(0) = 1$ is not possible. Since $a_2(x) = x$ is 0 at $x = 0$, Theorem 4.1 is not violated.

9. From $y = c_1 e^x \cos x + c_2 e^x \sin x$ we find

$$y' = c_1 e^x (-\sin x + \cos x) + c_2 e^x (\cos x + \sin x).$$

(a) We have $y(0) = c_1 = 1$, $y'(0) = c_1 + c_2 = 0$ so that $c_1 = 1$ and $c_2 = -1$. The solution is $y = e^x \cos x - e^x \sin x$.

(b) We have $y(0) = c_1 = 1$, $y(\pi) = -c_1 e^\pi = -1$, which is not possible.

(c) We have $y(0) = c_1 = 1$, $y(\pi/2) = c_2 e^{\pi/2} = 1$ so that $c_1 = 1$ and $c_2 = e^{-\pi/2}$. The solution is $y = e^x \cos x + e^{-\pi/2} e^x \sin x$.

(d) We have $y(0) = c_1 = 0$, $y(\pi) = -c_1 e^\pi = 0$ so that $c_1 = 0$ and c_2 is arbitrary. Solutions are $y = c_2 e^x \sin x$, for any real numbers c_2.

12. Since $a_0(x) = \tan x$ and $x_0 = 0$ the problem has a unique solution for $-\pi/2 < x < \pi/2$.

15. Since $(-4)x + (3)x^2 + (1)(4x - 3x^2) = 0$ the functions are linearly dependent.

18. Since $(1) \cos 2x + (1)1 + (-2) \cos^2 x = 0$ the functions are linearly dependent.

21. The functions are linearly independent since

$$W\left(1 + x, x, x^2\right) = \begin{vmatrix} 1 + x & x & x^2 \\ 1 & 1 & 2x \\ 0 & 0 & 2 \end{vmatrix} = 2 \neq 0.$$

24. $W\left(1 + x, x^3\right) = \begin{vmatrix} 1 + x & x^3 \\ 1 & 3x^2 \end{vmatrix} = x^2(3 + 2x) \neq 0$ for $-\infty < x < \infty$.

27. $W\left(e^x, e^{-x}, e^{4x}\right) = \begin{vmatrix} e^x & e^{-x} & e^{4x} \\ e^x & -e^{-x} & 4e^{4x} \\ e^x & e^{-x} & 16e^{4x} \end{vmatrix} = -30e^{4x} \neq 0$ for $-\infty < x < \infty$.

27

30. (a) The graphs of f_1 and f_2 are as shown. Obviously, neither function is a constant multiple of the other on $-\infty < x < \infty$. Hence, f_1 and f_2 are linearly independent on $(-\infty, \infty)$.

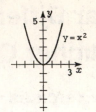 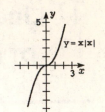

(b) For $x \geq 0$, $f_2 = x^2$ and so

$$W(f_1, f_2) = \begin{vmatrix} x^2 & x^2 \\ 2x & 2x \end{vmatrix} = 2x^3 - 2x^3 = 0.$$

For $x < 0$, $f_2 = -x^2$ and

$$W(f_1, f_2) = \begin{vmatrix} x^2 & -x^2 \\ 2x & -2x \end{vmatrix} = -2x^3 + 2x^3 = 0.$$

We conclude that $W(f_1, f_2) = 0$ for all real values of x.

33. The functions satisfy the differential equation and are linearly independent since

$$W\left(e^{-3x}, e^{4x}\right) = 7e^x \neq 0$$

for $-\infty < x < \infty$. The general solution is

$$y = c_1 e^{-3x} + c_2 e^{4x}.$$

36. The functions satisfy the differential equation and are linearly independent since

$$W\left(e^{x/2}, xe^{x/2}\right) = e^x \neq 0$$

for $-\infty < x < \infty$. The general solution is

$$y = c_1 e^{x/2} + c_2 x e^{x/2}.$$

39. The functions satisfy the differential equation and are linearly independent since

$$W\left(x, x^{-2}, x^{-2}\ln x\right) = 9x^{-6} \neq 0$$

for $0 < x < \infty$. The general solution is

$$y = c_1 x + c_2 x^{-2} + c_3 x^{-2}\ln x.$$

42. The functions $y_1 = \cos x$ and $y_2 = \sin x$ form a fundamental set of solutions of the homogeneous equation, and $y_p = x\sin x + (\cos x)\ln(\cos x)$ is a particular solution of the nonhomogeneous equation.

45. (a) From the graphs of $y_1 = x^3$ and $y_2 = |x|^3$ we see that the functions are linearly independent since they cannot be multiples of each other. It is easily shown that $y_1 = x^3$ solves $x^2 y'' - 4xy' + 6y = 0$. To show that $y_2 = |x|^3$ is a solution let $y_2 = x^3$ for $x \geq 0$ and let $y_2 = -x^3$ for $x < 0$.

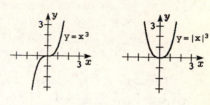

(b) If $x \geq 0$ then $y_2 = x^3$ and $W(y_1, y_2) = \begin{vmatrix} x^3 & x^3 \\ 3x^2 & 3x^2 \end{vmatrix} = 0$. If $x < 0$ then $y_2 = -x^3$ and

$$W(y_1, y_2) = \begin{vmatrix} x^3 & -x^3 \\ 3x^2 & -3x^2 \end{vmatrix} = 0.$$

(c) Part (b) does not violate Theorem 4.4 since $a_2(x) = x^2$ is zero at $x = 0$.

(d) The functions $Y_1 = x^3$ and $Y_2 = x^2$ are solutions of $x^2 y'' - 4xy' + 6y = 0$. They are linearly independent since $W\left(x^3, x^2\right) = x^4 \neq 0$ for $-\infty < x < \infty$.

(e) The function $y = x^3$ satisfies $y(0) = 0$ and $y'(0) = 0$.

(f) Neither is the general solution since we form a general solution on an interval for which $a_2(x) \neq 0$ for every x in the interval.

48. We identify $a_2(x) = 1 - x^2$ and $a_1(x) = -2x$. Then from Abel's formula in Problem 47 we have

$$W = ce^{-\int [a_1(x)/a_2(x)]\, dx} = ce^{-\int [-2x/(1-x^2)]\, dx} = ce^{-\ln(1-x^2)} = \frac{c}{1 - x^2}.$$

Exercises 4.2

In Problems 3-9 we use reduction of order to find a second solution. In Problems 12-30 we use formula (4) from the text.

3. Define $y = u(x)e^{2x}$ so

$$y' = 2ue^{2x} + u'e^{2x}, \quad y'' = e^{2x}u'' + 4e^{2x}u' + 4e^{2x}u, \quad \text{and} \quad y'' - 4y' + 4y = 4e^{2x}u'' = 0.$$

Therefore $u'' = 0$ and $u = c_1 x + c_2$. Taking $c_1 = 1$ and $c_2 = 0$ we see that a second solution is $y_2 = xe^{2x}$.

6. Define $y = u(x)\sin 3x$ so

$$y' = 3u\cos 3x + u'\sin 3x, \quad y'' = u''\sin 3x + 6u'\cos 3x - 9u\sin 3x,$$

and

$$y'' + 9y = (\sin 3x)u'' + 6(\cos 3x)u' = 0 \quad \text{or} \quad u'' + 9(\cot 3x)u' = 0.$$

If $w = u'$ we obtain the first-order equation $w' + 6(\cot 3x)w = 0$ which has the integrating factor $e^{6 \int \cot 3x\, dx} = \sin^2 3x$. Now

$$\frac{d}{dx}[(\sin^2 3x)w] = 0 \quad \text{gives} \quad (\sin^2 3x)w = c.$$

Therefore $w = u' = c \csc^2 3x$ and $u = c_1 \cot 3x$. A second solution is $y_2 = \cot 3x \sin 3x = \cos 3x$.

9. Define $y = u(x)e^{2x/3}$ so

$$y' = \frac{2}{3}e^{2x/3}u + e^{2x/3}u', \quad y'' = e^{2x/3}u'' + \frac{4}{3}e^{2x/3}u' + \frac{4}{9}e^{2x/3}u$$

and

$$9y'' - 12y' + 4y = 9e^{2x/3}u'' = 0.$$

Therefore $u'' = 0$ and $u = c_1 x + c_2$. Taking $c_1 = 1$ and $c_2 = 0$ we see that a second solution is $y_2 = xe^{2x/3}$.

12. Identifying $P(x) = 2/x$ we have

$$y_2 = x^2 \int \frac{e^{-\int (2/x)\, dx}}{x^4}\, dx = x^2 \int x^{-6} dx = -\frac{1}{5}x^{-3}.$$

A second solution is $y_2 = x^{-3}$.

15. Identifying $P(x) = 2(1+x)/\left(1 - 2x - x^2\right)$ we have

$$y_2 = (x+1) \int \frac{e^{-\int 2(1+x)dx/(1-2x-x^2)}}{(x+1)^2}\, dx = (x+1) \int \frac{e^{\ln(1-2x-x^2)}}{(x+1)^2}\, dx$$

$$= (x+1) \int \frac{1 - 2x - x^2}{(x+1)^2}\, dx = (x+1) \int \left[\frac{2}{(x+1)^2} - 1\right] dx$$

$$= (x+1) \left[-\frac{2}{x+1} - x\right] = -2 - x^2 - x.$$

A second solution is $y_2 = x^2 + x + 2$.

18. Identifying $P(x) = -3/x$ we have

$$y_2 = x^2 \cos(\ln x) \int \frac{e^{-\int -3\, dx/x}}{x^4 \cos^2(\ln x)}\, dx = x^2 \cos(\ln x) \int \frac{x^3}{x^4 \cos^2(\ln x)}\, dx$$

$$= x^2 \cos(\ln x)\tan(\ln x) = x^2 \sin(\ln x).$$

A second solution is $y_2 = x^2 \sin(\ln x)$.

21. Identifying $P(x) = -1/x$ we have

$$y_2 = x \int \frac{e^{-\int -dx/x}}{x^2}\, dx = x \int \frac{dx}{x} = x \ln|x|.$$

A second solution is $y_2 = x \ln|x|$.

24. Identifying $P(x) = 1/x$ we have

$$y_2 = \cos(\ln x) \int \frac{e^{-\int dx/x}}{\cos^2(\ln x)} \, dx = \cos(\ln x) \int \frac{1/x}{\cos^2(\ln x)} \, dx = \cos(\ln x)\tan(\ln x) = \sin(\ln x).$$

A second solution is $y_2 = \sin(\ln x)$.

27. Identifying $P(x) = -(9x + 6)/(3x + 1)$ we have

$$y_2 = e^{3x} \int \frac{e^{-\int -(9x+6)dx/(3x+1)}}{e^{6x}} \, dx = e^{3x} \int \frac{e^{\int [3+3/(3x+1)]dx}}{e^{6x}} \, dx = e^{3x} \int \frac{e^{3x+\ln(3x+1)}}{e^{6x}} \, dx$$

$$= e^{3x} \int \frac{(3x+1)e^{3x}}{e^{6x}} \, dx = e^{3x} \int (3x+1)e^{-3x} \, dx = e^{3x} \left(-xe^{-3x} - \frac{2}{3}e^{-3x} \right) = -x - \frac{2}{3}.$$

A second solution is $y_2 = 3x + 2$.

30. Identifying $P(x) = -(2+x)/x$ we have

$$y_2 = \int e^{-\int -(2+x)dx/x} \, dx = \int e^{2\ln x + x} \, dx = \int x^2 e^x \, dx = \left(x^2 - 2x + 2 \right) e^x.$$

A second solution is $y_2 = \left(x^2 - 2x + 2 \right) e^x$.

33. Identifying $P(x) = -3$ we have

$$y_2 = e^x \int \frac{e^{-\int -3\,dx}}{e^{2x}} \, dx = e^x \int e^x \, dx = e^{2x}.$$

To find a particular solution we try $y_p = Ae^{3x}$. Then $y' = 3Ae^{3x}$, $y'' = 9Ae^{3x}$, and $9Ae^{3x} - 3\left(3Ae^{3x} \right) + 2Ae^{3x} = 5e^{3x}$. Thus $A = 5/2$ and $y_p = \frac{5}{2}e^{3x}$. The general solution is

$$y = c_1 e^x + c_2 e^{2x} + \frac{5}{2}e^{3x}.$$

Exercises 4.3

3. From $m^2 - 36 = 0$ we obtain $m = 6$ and $m = -6$ so that

$$y = c_1 e^{6x} + c_2 e^{-6x}.$$

6. From $3m^2 + 1 = 0$ we obtain $m = i/\sqrt{3}$ and $m = -i/\sqrt{3}$ so that

$$y = c_1 \cos x/\sqrt{3} + c_2 \sin x/\sqrt{3}.$$

9. From $m^2 + 8m + 16 = 0$ we obtain $m = -4$ and $m = -4$ so that

$$y = c_1 e^{-4x} + c_2 x e^{-4x}.$$

12. From $m^2 + 4m - 1 = 0$ we obtain $m = -2 \pm \sqrt{5}$ so that
$$y = c_1 e^{(-2+\sqrt{5})x} + c_2 e^{(-2-\sqrt{5})x}.$$

15. From $m^2 - 4m + 5 = 0$ we obtain $m = 2 \pm i$ so that
$$y = e^{2x}(c_1 \cos x + c_2 \sin x).$$

18. From $2m^2 + 2m + 1 = 0$ we obtain $m = -1/2 \pm i/2$ so that
$$y = e^{-x/2}(c_1 \cos x/2 + c_2 \sin x/2).$$

21. From $m^3 - 1 = 0$ we obtain $m = 1$ and $m = -1/2 \pm \sqrt{3}\,i/2$ so that
$$y = c_1 e^x + e^{-x/2}\left(c_2 \cos \sqrt{3}\,x/2 + c_3 \sin \sqrt{3}\,x/2\right).$$

24. From $m^3 + 3m^2 - 4m - 12 = 0$ we obtain $m = -2$, $m = 2$, and $m = -3$ so that
$$y = c_1 e^{-2x} + c_2 e^{2x} + c_3 e^{-3x}.$$

27. From $m^3 + 3m^2 + 3m + 1 = 0$ we obtain $m = -1$, $m = -1$, and $m = -1$ so that
$$y = c_1 e^{-x} + c_2 x e^{-x} + c_3 x^2 e^{-x}.$$

30. From $m^4 - 2m^2 + 1 = 0$ we obtain $m = 1$, $m = 1$, $m = -1$, and $m = -1$ so that
$$y = c_1 e^x + c_2 x e^x + c_3 e^{-x} + c_4 x e^{-x}.$$

33. From $m^5 - 16m = 0$ we obtain $m = 0$, $m = 2$, $m = -2$, and $m = \pm 2i$ so that
$$y = c_1 + c_2 e^{2x} + c_3 e^{-2x} + c_4 \cos 2x + c_5 \sin 2x.$$

36. From $2m^5 - 7m^4 + 12m^3 + 8m^2 = 0$ we obtain $m = 0$, $m = 0$, $m = -1/2$, and $m = 2 \pm 2i$ so that
$$y = c_1 + c_2 x + c_3 e^{-x/2} + e^{2x}(c_4 \cos 2x + c_5 \sin 2x).$$

39. From $m^2 + 6m + 5 = 0$ we obtain $m = -1$ and $m = -5$ so that $y = c_1 e^{-x} + c_2 e^{-5x}$. If $y(0) = 0$ and $y'(0) = 3$ then $c_1 + c_2 = 0$, $-c_1 - 5c_2 = 3$, so $c_1 = 3/4$, $c_2 = -3/4$, and
$$y = \frac{3}{4}e^{-x} - \frac{3}{4}e^{-5x}.$$

42. From $m^2 - 2m + 1 = 0$ we obtain $m = 1$ and $m = 1$ so that $y = c_1 e^x + c_2 x e^x$. If $y(0) = 5$ and $y'(0) = 10$ then $c_1 = 5$, $c_1 + c_2 = 10$ so $c_1 = 5$, $c_2 = 5$, and
$$y = 5e^x + 5x e^x.$$

45. From $m^2 - 3m + 2 = 0$ we obtain $m = 1$ and $m = 2$ so that $y = c_1 e^x + c_2 e^{2x}$. If $y(1) = 0$ and $y'(1) = 1$ then $c_1 e + c_2 e^2 = 0$, $c_1 e + 2c_2 e^2 = 0$ so $c_1 = -e^{-1}$, $c_2 = e^{-2}$, and
$$y = -e^{x-1} + e^{2x-2}.$$

48. From $m^3 + 2m^2 - 5m - 6 = 0$ we obtain $m = -1$, $m = 2$, and $m = -3$ so that

$$y = c_1 e^{-x} + c_2 e^{2x} + c_3 e^{-3x}.$$

If $y(0) = 0$, $y'(0) = 0$, and $y''(0) = 1$ then

$$c_1 + c_2 + c_3 = 0, \quad -c_1 + 2c_2 - 3c_3 = 0, \quad c_1 + 4c_2 + 9c_3 = 1,$$

so $c_1 = -1/6$, $c_2 = 1/15$, $c_3 = 1/10$, and

$$y = -\frac{1}{6}e^{-x} + \frac{1}{15}e^{2x} + \frac{1}{10}e^{-3x}.$$

51. From $m^4 - 3m^3 + 3m^2 - m = 0$ we obtain $m = 0$, $m = 1$, $m = 1$, and $m = 1$ so that $y = c_1 + c_2 e^x + c_3 x e^x + c_4 x^2 e^x$. If $y(0) = 0$, $y'(0) = 0$, $y''(0) = 1$, and $y'''(0) = 1$ then

$$c_1 + c_2 = 0, \quad c_2 + c_3 = 0, \quad c_2 + 2c_3 + 2c_4 = 1, \quad c_2 + 3c_3 + 6c_4 = 1,$$

so $c_1 = 2$, $c_2 = -2$, $c_3 = 2$, $c_4 = -1/2$, and

$$y = 2 - 2e^x + 2xe^x - \frac{1}{2}x^2 e^x.$$

54. From $m^2 + 4 = 0$ we obtain $m = \pm 2i$ so that $y = c_1 \cos 2x + c_2 \sin 2x$. If $y(0) = 0$ and $y(\pi) = 0$ then $c_1 = 0$ and $y = c_2 \sin 2x$.

57. Since $(m - 4)(m + 5)^2 = m^3 + 6m^2 - 15m - 100$ the differential equation is

$$y''' + 6y'' - 15y' - 100y = 0.$$

60. From the solution $y_1 = e^{-4x} \cos x$ we conclude that $m_1 = -4 + i$ and $m_2 = -4 - i$ are roots of the auxiliary equation. Hence another solution must be $y_2 = e^{-4x} \sin x$. Now dividing the polynomial $m^3 + 6m^2 + m - 34$ by $[m - (-4+i)][m - (-4-i)] = m^2 + 8m + 17$ gives $m - 2$. Therefore $m_3 = 2$ is the third root of the auxiliary equation, and the general solution of the differential equation is

$$y = c_1 e^{-4x} \cos x + c_2 e^{-4x} \sin x + c_3 e^{2x}.$$

63. Since $m^2(m - 7) = m^3 - 7m^2$, a differential equation is

$$y''' - 7y'' = 0.$$

Exercises 4.4

3. From $m^2 - 10m + 25 = 0$ we find $m_1 = m_2 = 5$. Then $y_c = c_1 e^{5x} + c_2 x e^{5x}$ and we assume $y_p = Ax + B$. Substituting into the differential equation we obtain $25A = 30$ and $-10A + 25B = 3$. Then $A = \frac{6}{5}$, $B = \frac{6}{5}$, $y_p = \frac{6}{5}x + \frac{6}{5}$, and

$$y = c_1 e^{5x} + c_2 x e^{5x} + \frac{6}{5}x + \frac{6}{5}.$$

6. From $m^2 - 8m + 20 = 0$ we find $m_1 = 2 + 4i$ and $m_2 = 2 - 4i$. Then $y_c = e^{2x}(c_1 \cos 4x + c_2 \sin 4x)$ and we assume $y_p = Ax^2 + Bx + C + (Dx + E)e^x$. Substituting into the differential equation we obtain

$$2A - 8B + 20C = 0$$

$$-6D + 13E = 0$$

$$-16A + 20B = 0$$

$$13D = -26$$

$$20A = 100.$$

Then $A = 5$, $B = 4$, $C = \frac{11}{10}$, $D = -2$, $E = -\frac{12}{13}$, $y_p = 5x^2 + 4x + \frac{11}{10} + \left(-2x - \frac{12}{13}\right)e^x$ and

$$y = e^{2x}(c_1 \cos 4x + c_2 \sin 4x) + 5x^2 + 4x + \frac{11}{10} + \left(-2x - \frac{12}{13}\right)e^x.$$

9. From $m^2 - m = 0$ we find $m_1 = 1$ and $m_2 = 0$. Then $y_c = c_1 e^x + c_2$ and we assume $y_p = Ax$. Substituting into the differential equation we obtain $-A = -3$. Then $A = 3$, $y_p = 3x$ and $y = c_1 e^x + c_2 + 3x$.

12. From $m^2 - 16 = 0$ we find $m_1 = 4$ and $m_2 = -4$. Then $y_c = c_1 e^{4x} + c_2 e^{-4x}$ and we assume $y_p = Axe^{4x}$. Substituting into the differential equation we obtain $8A = 2$. Then $A = \frac{1}{4}$, $y_p = \frac{1}{4}xe^{4x}$ and

$$y = c_1 e^{4x} + c_2 e^{-4x} + \frac{1}{4}xe^{4x}.$$

15. From $m^2 + 1 = 0$ we find $m_1 = i$ and $m_2 = -i$. Then $y_c = c_1 \cos x + c_2 \sin x$ and we assume $y_p = (Ax^2 + Bx)\cos x + (Cx^2 + Dx)\sin x$. Substituting into the differential equation we obtain $4C = 0$, $2A + 2D = 0$, $-4A = 2$, and $-2B + 2C = 0$. Then $A = -\frac{1}{2}$, $B = 0$, $C = 0$, $D = \frac{1}{2}$, $y_p = -\frac{1}{2}x^2 \cos x + \frac{1}{2}x \sin x$, and

$$y = c_1 \cos x + c_2 \sin x - \frac{1}{2}x^2 \cos x + \frac{1}{2}x \sin x.$$

18. From $m^2 - 2m + 2 = 0$ we find $m_1 = 1 + i$ and $m_2 = 1 - i$. Then $y_c = e^x(c_1 \cos x + c_2 \sin x)$ and we assume $y_p = Ae^{2x} \cos x + Be^{2x} \sin x$. Substituting into the differential equation we obtain $A + 2B = 1$ and $-2A + B = -3$. Then $A = \frac{7}{5}$, $B = -\frac{1}{5}$, $y_p = \frac{7}{5}e^{2x} \cos x - \frac{1}{5}e^{2x} \sin x$ and

$$y = e^x(c_1 \cos x + c_2 \sin x) + \frac{7}{5}e^{2x} \cos x - \frac{1}{5}e^{2x} \sin x.$$

21. From $m^3 - 6m^2 = 0$ we find $m_1 = m_2 = 0$ and $m_3 = 6$. Then $y_c = c_1 + c_2 x + c_3 e^{6x}$ and we assume $y_p = Ax^2 + B \cos x + C \sin x$. Substituting into the differential equation we obtain $-12A = 3$, $6B - C = -1$, and $B + 6C = 0$. Then $A = -\frac{1}{4}$, $B = -\frac{6}{37}$, $C = \frac{1}{37}$, $y_p = -\frac{1}{4}x^2 - \frac{6}{37} \cos x + \frac{1}{37} \sin x$, and

$$y = c_1 + c_2 x + c_3 e^{6x} - \frac{1}{4}x^2 - \frac{6}{37} \cos x + \frac{1}{37} \sin x.$$

24. From $m^3 - m^2 - 4m + 4 = 0$ we find $m_1 = 1$, $m_2 = 2$, and $m_3 = -2$. Then $y_c = c_1 e^x + c_2 e^{2x} + c_3 e^{-2x}$ and we assume $y_p = A + Bxe^x + Cxe^{2x}$. Substituting into the differential equation we obtain $4A = 5$, $-3B = -1$, and $4C = 1$. Then $A = \frac{5}{4}$, $B = \frac{1}{3}$, $C = \frac{1}{4}$, $y_p = \frac{5}{4} + \frac{1}{3}xe^x + \frac{1}{4}xe^{2x}$, and

$$y = c_1 e^x + c_2 e^{2x} + c_3 e^{-2x} + \frac{5}{4} + \frac{1}{3}xe^x + \frac{1}{4}xe^{2x}.$$

27. We write $8 \sin^2 x = 4 - 4 \cos 2x$. From $m^2 + 1 = 0$ we find $m_1 = i$ and $m_2 = -i$. Then $y_c = c_1 \cos x + c_2 \sin x$ and we assume $y_p = A + B \cos 2x + C \sin 2x$. Substituting into the differential equation we obtain $A = 4$, $-3B = -4$, and $-3C = 0$. Then $A = 4$, $B = \frac{4}{3}$, $C = 0$, and $y_p = 4 + \frac{4}{3} \cos 2x$.

30. We have $y_c = c_1 e^{-2x} + c_2 e^{x/2}$ and we assume $y_p = Ax^2 + Bx + C$. Substituting into the differential equation we find $A = -7$, $B = -19$, and $C = -37$. Thus $y = c_1 e^{-2x} + c_2 e^{x/2} - 7x^2 - 19x - 37$. From the initial conditions we obtain $c_1 = -\frac{1}{5}$ and $c_2 = \frac{186}{5}$, so

$$y = -\frac{1}{5}e^{-2x} + \frac{186}{5}e^{x/2} - 7x^2 - 19x - 37.$$

33. We have $y_c = e^{-2x}(c_1 \cos x + c_2 \sin x)$ and we assume $y_p = Ae^{-4x}$. Substituting into the differential equation we find $A = 5$. Thus $y = e^{-2x}(c_1 \cos x + c_2 \sin x) + 7e^{-4x}$. From the initial conditions we obtain $c_1 = -10$ and $c_2 = 9$, so

$$y = e^{-2x}(-10 \cos x + 9 \sin x + 7e^{-4x}).$$

36. We have $x_c = c_1 \cos \omega t + c_2 \sin \omega t$ and we assume $x_p = A \cos \gamma t + B \sin \gamma t, \gamma \neq \omega$. Substituting into the differential equation we find $A = F_0/(\omega^2 - \gamma^2)$ and $B = 0$. Thus

$$x = c_1 \cos \omega t + c_2 \sin \omega t + \frac{F_0}{(\omega^2 - \gamma^2)} \cos \gamma t.$$

From the initial conditions we obtain $c_1 = -F_0/(\omega^2 - \gamma^2)$ and $c_2 = 0$, so

$$x = -\frac{F_0}{(\omega^2 - \gamma^2)} \cos \omega t + \frac{F_0}{(\omega^2 - \gamma^2)} \cos \gamma t.$$

39. We have $y_c = c_1 + c_2 e^x + c_3 x e^x$ and we assume $y_p = Ax + Bx^2 e^x + Ce^{5x}$. Substituting into the differential equation we find $A = 2$, $B = -12$, and $C = \frac{1}{2}$. Thus

$$y = c_1 + c_2 e^x + c_3 x e^x + 2x - 12x^2 e^x + \frac{1}{2}e^{5x}.$$

From the initial conditions we obtain $c_1 = 11$, $c_2 = -11$, and $c_3 = 9$, so

$$y = 11 - 11e^x + 9xe^x + 2x - 12x^2 e^x + \frac{1}{2}e^{5x}.$$

42. We have $y_c = e^x(c_1 \cos x + c_2 \sin x)$ and we assume $y_p = Ax + B$. Substituting into the differential equation we find $A = 1$ and $B = 0$. Thus $y = e^x(c_1 \cos x + c_2 \sin x) + x$. From $y(0) = 0$ and $y(\pi) = \pi$ we obtain

$$c_1 = 0$$

$$\pi - e^\pi c_1 = \pi.$$

Solving this system we find $c_1 = 0$ and c_2 is any real number. The solution of the boundary-value problem is

$$y = c_2 e^x \sin x + x.$$

Exercises 4.5

3. $(3D^2 - 5D + 1)y = e^x$

6. $(D^4 - 2D^2 + D) = e^{-3x} + e^{2x}$

9. $D^2 - 4D - 12 = (D - 6)(D + 2)$

12. $D^3 + 4D = D(D^2 + 4)$

15. $D^4 + 8D = D(D + 2)(D^2 - 2D + 4)$

18. $(2D - 1)y = (2D - 1)4e^{x/2} = 8De^{x/2} - 4e^{x/2} = 4e^{x/2} - 4e^{x/2} = 0$

21. D^4 because of x^3

24. $D^2(D - 6)^2$ because of x and xe^{6x}

27. $D^3(D^2 + 16)$ because of x^2 and $\sin 4x$

30. $D(D - 1)(D - 2)$ because of 1, e^x, and e^{2x}

33. $1, x, x^2, x^3, x^4$

36. $D^2 - 9D - 36 = (D - 12)(D + 3);\quad e^{12x},\ e^{-3x}$

39. $D^3 - 10D^2 + 25D = D(D - 5)^2;\quad 1,\ e^{5x},\ xe^{5x}$

_____ Exercises 4.6 _____

3. Applying D to the differential equation we obtain

$$D(D^2 + D)y = D^2(D + 1)y = 0.$$

Then

$$y = \underbrace{c_1 + c_2 e^{-x}}_{y_c} + c_3 x$$

and $y_p = Ax$. Substituting y_p into the differential equation yields $A = 3$. The general solution is

$$y = c_1 + c_2 e^{-3x} + 3x.$$

6. Applying D^2 to the differential equation we obtain

$$D^2(D^2 + 3D)y = D^3(D + 3)y = 0.$$

Then

$$y = \underbrace{c_1 + c_2 e^{-3x}}_{y_c} + c_3 x^2 + c_4 x$$

and $y_p = Ax^2 + Bx$. Substituting y_p into the differential equation yields $6Ax + (2A + 3B) = 4x - 5$. Equating coefficients gives

$$6A = 4$$

$$2A + 3B = -5.$$

Then $A = 2/3$, $B = -19/9$, and the general solution is

$$y = c_1 + c_2 e^{-3x} + \frac{2}{3}x^2 - \frac{19}{9}x.$$

9. Applying $D - 4$ to the differential equation we obtain

$$(D - 4)(D^2 - D - 12)y = (D - 4)^2(D + 3)y = 0.$$

Then

$$y = \underbrace{c_1 e^{4x} + c_2 e^{-3x}}_{y_c} + c_3 x e^{4x}$$

and $y_p = Axe^{4x}$. Substituting y_p into the differential equation yields $7Ae^{4x} = e^{4x}$. Equating coefficients gives $A = 1/7$. The general solution is

$$y = c_1 e^{4x} + c_2 e^{-3x} + \frac{1}{7}xe^{4x}.$$

37

12. Applying $D^2(D+2)$ to the differential equation we obtain

$$D^2(D+2)(D^2+6D+8)y = D^2(D+2)^2(D+4)y = 0.$$

Then

$$y = \underbrace{c_1e^{-2x} + c_2e^{-4x}}_{y_c} + c_3xe^{-2x} + c_4x + c_5$$

and $y_p = Axe^{-2x} + Bx + C$. Substituting y_p into the differential equation yields $2Ae^{-2x} + 8Bx + (6B+8C) = 3e^{-2x} + 2x$. Equating coefficients gives

$$2A = 3$$

$$8B = 2$$

$$6B + 8C = 0.$$

Then $A = 3/2$, $B = 1/4$, $C = -3/16$, and the general solution is

$$y = c_1e^{-2x} + c_2e^{-4x} + \frac{3}{2}xe^{-2x} + \frac{1}{4}x - \frac{3}{16}.$$

15. Applying $(D-4)^2$ to the differential equation we obtain

$$(D-4)^2(D^2+6D+9)y = (D-4)^2(D+3)^2y = 0.$$

Then

$$y = \underbrace{c_1e^{-3x} + c_2xe^{-3x}}_{y_c} + c_3xe^{4x} + c_4e^{4x}$$

and $y_p = Axe^{4x} + Be^{4x}$. Substituting y_p into the differential equation yields $49Axe^{4x} + (14A+49B)e^{4x} = -xe^{4x}$. Equating coefficients gives

$$49A = -1$$

$$14A + 49B = 0.$$

Then $A = -1/49$, $B = 2/343$, and the general solution is

$$y = c_1e^{-3x} + c_2xe^{-3x} - \frac{1}{49}xe^{4x} + \frac{2}{343}e^{4x}.$$

18. Applying $(D+1)^3$ to the differential equation we obtain

$$(D+1)^3(D^2+2D+1)y = (D+1)^5y = 0.$$

Then

$$y = \underbrace{c_1e^{-x} + c_2xe^{-x}}_{y_c} + c_3x^4e^{-x} + c_4x^3e^{-x} + c_5x^2e^{-x}$$

and $y_p = Ax^4e^{-x} + Bx^3e^{-x} + Cx^2e^{-x}$. Substituting y_p into the differential equation yields $12Ax^2e^{-x} + 6Bxe^{-x} + 2Ce^{-x} = x^2e^{-x}$. Equating coefficients gives $A = 1/12$, $B = 0$, and $C = 0$. The general solution is

$$y = c_1e^{-x} + c_2xe^{-x} + \frac{1}{2}x^4e^{-x}.$$

21. Applying $D^2 + 25$ to the differential equation we obtain

$$(D^2 + 25)(D^2 + 25) = (D^2 + 25)^2 = 0.$$

Then

$$y = \underbrace{c_1 \cos 5x + c_2 \sin 5x}_{y_c} + c_3x \cos 5x + c_4x \cos 5x$$

and $y_p = Ax \cos 5x + Bx \sin 5x$. Substituting y_p into the differential equation yields $10B \cos 5x - 10A \sin 5x = 20 \sin 5x$. Equating coefficients gives $A = -2$ and $B = 0$. The general solution is

$$y = c_1 \cos 5x + c_2 \sin 5x - 2x \cos 5x.$$

24. Writing $\cos^2 x = \frac{1}{2}(1 + \cos 2x)$ and applying $D(D^2 + 4)$ to the differential equation we obtain

$$D(D^2 + 4)(D^2 + 4) = D(D^2 + 4)^2 = 0.$$

Then

$$y = \underbrace{c_1 \cos 2x + c_2 \sin 2x}_{y_c} + c_3x \cos 2x + c_4x \sin 2x + c_5$$

and $y_p = Ax \cos 2x + Bx \sin 2x + C$. Substituting y_p into the differential equation yields $-4A \sin 2x + 4B \cos 2x + 4C = \frac{1}{2} + \frac{1}{2} \cos 2x$. Equating coefficients gives $A = 0$, $B = 1/8$, and $C = 1/8$. The general solution is

$$y = c_1 \cos 2x + c_2 \sin 2x + \frac{1}{8}x \sin 2x + \frac{1}{8}.$$

27. Applying $D^2(D - 1)$ to the differential equation we obtain

$$D^2(D - 1)(D^3 - 3D^2 + 3D - 1) = D^2(D - 1)^4 = 0.$$

Then

$$y = \underbrace{c_1e^x + c_2xe^x + c_3x^2e^x}_{y_c} + c_4 + c_5x + c_6x^3e^x$$

and $y_p = A + Bx + Cx^3e^x$. Substituting y_p into the differential equation yields $(-A + 3B) - Bx + 6Ce^x = 16 - x + e^x$. Equating coefficients gives

$$-A + 3B = 16$$

$$-B = -1$$

$$6C = 1.$$

Then $A = -13$, $B = 1$, and $C = 1/6$, and the general solution is

$$y = c_1 e^x + c_2 x e^x + c_3 x^2 e^x - 13 + x + \frac{1}{6} x^3 e^x.$$

30. Applying $D^3(D-2)$ to the differential equation we obtain

$$D^3(D-2)(D^4 - 4D^2) = D^5(D-2)^2(D+2) = 0.$$

Then

$$y = \underbrace{c_1 + c_2 x + c_3 e^{2x} + c_4 e^{-2x}}_{y_c} + c_5 x^2 + c_6 x^3 + c_7 x^4 + c_8 x e^{2x}$$

and $y_p = Ax^2 + Bx^3 + Cx^4 + Dxe^{2x}$. Substituting y_p into the differential equation yields $(-8A + 24C) - 24Bx - 48Cx^2 + 16De^{2x} = 5x^2 - e^{2x}$. Equating coefficients gives

$$-8A + 24C = 0$$

$$-24B = 0$$

$$-48C = 5$$

$$16D = -1.$$

Then $A = -5/16$, $B = 0$, $C = -5/48$, and $D = -1/16$, and the general solution is

$$y = c_1 + c_2 x + c_3 e^{2x} + c_4 e^{-2x} - \frac{5}{16} x^2 - \frac{5}{48} x^4 - \frac{1}{16} x e^{2x}.$$

33. The complementary function is $y_c = c_1 e^{8x} + c_2 e^{-8x}$. Using D to annihilate 16 we find $y_p = A$. Substituting y_p into the differential equation we obtain $-64A = 16$. Thus $A = -1/4$ and

$$y = c_1 e^{8x} + c_2 e^{-8x} - \frac{1}{4}$$

$$y' = 8c_1 e^{8x} - 8c_2 e^{-8x}.$$

The initial conditions imply

$$c_1 + c_2 = \frac{5}{4}$$

$$8c_1 - 8c_2 = 0.$$

Thus $c_1 = c_2 = 5/8$ and

$$y = \frac{5}{8} e^{8x} + \frac{5}{8} e^{-8x} - \frac{1}{4}.$$

36. The complementary function is $y_c = c_1 e^x + c_2 e^{-6x}$. Using $D - 2$ to annihilate $10e^{2x}$ we find $y_p = Ae^{2x}$. Substituting y_p into the differential equation we obtain $8Ae^{2x} = 10e^{2x}$. Thus $A = 5/4$

and

$$y = c_1 e^x + c_2 e^{-6x} + \frac{5}{4}e^{2x}$$

$$y' = c_1 e^x - 6c_2 e^{-6x} + \frac{5}{2}e^{2x}.$$

The initial conditions imply

$$c_1 + c_2 = -\frac{1}{4}$$

$$c_1 - 6c_2 = -\frac{3}{2}.$$

Thus $c_1 = -3/7$ and $c_2 = 5/28$, and

$$y = -\frac{3}{7}e^x + \frac{5}{28}e^{-6x} + \frac{5}{4}e^{2x}.$$

39. The complementary function is $y_c = e^{2x}(c_1 \cos 2x + c_2 \sin 2x)$. Using D^4 to annihilate x^3 we find $y_p = A + Bx + Cx^2 + Dx^3$. Substituting y_p into the differential equation we obtain $(8A - 4B + 2C) + (8B - 8C + 6D)x + (8C - 12D)x^2 + 8Dx^3 = x^3$. Thus $A = 0$, $B = 3/32$, $C = 3/16$, and $D = 1/8$, and

$$y = e^{2x}(c_1 \cos 2x + c_2 \sin 2x) + \frac{3}{32}x + \frac{3}{16}x^2 + \frac{1}{8}x^3$$

$$y' = e^{2x}\left[c_1(2\cos 2x - 2\sin 2x) + c_2(2\cos 2x + 2\sin 2x)\right] + \frac{3}{32} + \frac{3}{8}x + \frac{3}{8}x^2.$$

The initial conditions imply

$$c_1 = 2$$

$$2c_1 + 2c_2 + \frac{3}{32} = 4.$$

Thus $c_1 = 2$, $c_2 = -3/64$, and

$$y = e^{2x}\left(2\cos 2x - \frac{3}{64}\sin 2x\right) + \frac{3}{32}x + \frac{3}{16}x^2 + \frac{1}{8}x^3.$$

42. The complementary function is $y_c = c_1 + c_2 e^{-x}$. Using $D(D+1)(D^2+1)^3$ to annihilate $9 - e^{-x} + x^2 \sin x$ we obtain

$$y_p = Ax + Bxe^{-x} + C\cos x + D\sin x + Ex\cos x + Fx\sin x + Gx^2\cos x + Hx^2\sin x.$$

41

Exercises 4.7

The particular solution, $y_p = u_1 y_1 + u_2 y_2$, in the following problems can take on a variety of forms, especially where trigonometric functions are involved. The validity of a particular form can best be checked by substituting it back into the differential equation.

3. The auxiliary equation is $m^2 + 1 = 0$, so $y_c = c_1 \cos x + c_2 \sin x$ and

$$W = \begin{vmatrix} \cos x & \sin x \\ -\sin x & \cos x \end{vmatrix} = 1.$$

Identifying $f(x) = \sin x$ we obtain

$$u_1' = -\sin^2 x$$

$$u_2' = \cos x \sin x.$$

Then

$$u_1 = \frac{1}{4} \sin 2x - \frac{1}{2} x = \frac{1}{2} \sin x \cos x - \frac{1}{2} x$$

$$u_2 = -\frac{1}{2} \cos^2 x.$$

and

$$y = c_1 \cos x + c_2 \sin x + \frac{1}{2} \sin x \cos^2 x - \frac{1}{2} x \cos x - \frac{1}{2} \cos^2 x \sin x$$

$$= c_1 \cos x + c_2 \sin x - \frac{1}{2} x \cos x$$

for $-\infty < x < \infty$.

6. The auxiliary equation is $m^2 + 1 = 0$, so $y_c = c_1 \cos x + c_2 \sin x$ and

$$W = \begin{vmatrix} \cos x & \sin x \\ -\sin x & \cos x \end{vmatrix} = 1.$$

Identifying $f(x) = \sec^2 x$ we obtain

$$u_1' = -\frac{\sin x}{\cos^2 x}$$

$$u_2' = \sec x.$$

Then

$$u_1 = -\frac{1}{\cos x} = -\sec x$$

$$u_2 = \ln|\sec x + \tan x|$$

and

$$y = c_1 \cos x + c_2 \sin x - \cos x \sec x + \sin x \ln|\sec x + \tan x|$$

$$= c_1 \cos x + c_2 \sin x - 1 + \sin x \ln|\sec x + \tan x|$$

for $-\pi/2 < x < \pi/2$.

9. The auxiliary equation is $m^2 - 4 = 0$, so $y_c = c_1 e^{2x} + c_2 e^{-2x}$ and

$$W = \begin{vmatrix} e^{2x} & e^{-2x} \\ 2e^{2x} & -2e^{-2x} \end{vmatrix} = -4.$$

Identifying $f(x) = e^{2x}/x$ we obtain $u_1' = 1/4x$ and $u_2' = -e^{4x}/4x$. Then

$$u_1 = \frac{1}{4} \ln|x|, \qquad u_2 = -\frac{1}{4} \int_{x_0}^{x} \frac{e^{4t}}{t} \, dt$$

and

$$y = c_1 e^{2x} + c_2 e^{-2x} + \frac{1}{4} \left(e^{2x} \ln|x| - e^{-2x} \int_{x_0}^{x} \frac{e^{4t}}{t} \, dt \right), \qquad x_0 > 0$$

for $x > 0$.

12. The auxiliary equation is $m^2 - 3m + 2 = (m-1)(m-2) = 0$, so $y_c = c_1 e^x + c_2 e^{2x}$ and

$$W = \begin{vmatrix} e^x & e^{2x} \\ e^x & 2e^{2x} \end{vmatrix} = e^{3x}.$$

Identifying $f(x) = e^{3x}/(1 + e^x)$ we obtain

$$u_1' = -\frac{e^{2x}}{1 + e^x} = \frac{e^x}{1 + e^x} - e^x$$

$$u_2' = \frac{e^x}{1 + e^x}.$$

Then $u_1 = \ln(1 + e^x) - e^x$, $u_2 = \ln(1 + e^x)$, and

$$y = c_1 e^x + c_2 e^{2x} + e^x \ln(1 + e^x) - e^{2x} + e^{2x} \ln(1 + e^x)$$

$$= c_1 e^x + c_3 e^{2x} + (1 + e^x) e^x \ln(1 + e^x)$$

for $-\infty < x < \infty$.

15. The auxiliary equation is $m^2 - 2m + 1 = (m-1)^2 = 0$, so $y_c = c_1 e^x + c_2 x e^x$ and

$$W = \begin{vmatrix} e^x & x e^x \\ e^x & x e^x + e^x \end{vmatrix} = e^{2x}.$$

43

Identifying $f(x) = e^x / (1 + x^2)$ we obtain

$$u_1' = -\frac{xe^x e^x}{e^{2x}(1+x^2)} = -\frac{x}{1+x^2}$$

$$u_2' = \frac{e^x e^x}{e^{2x}(1+x^2)} = \frac{1}{1+x^2}.$$

Then $u_1 = -\frac{1}{2}\ln\left(1+x^2\right)$, $u_2 = \tan^{-1} x$, and

$$y = c_1 e^x + c_2 x e^x - \frac{1}{2}e^x \ln\left(1+x^2\right) + xe^x \tan^{-1} x$$

for $-\infty < x < \infty$.

18. The auxiliary equation is $m^2 + 10m + 25 = (m+5)^2 = 0$, so $y_c = c_1 e^{-5x} + c_2 x e^{-5x}$ and

$$W = \begin{vmatrix} e^{-5x} & xe^{-5x} \\ -5e^{-5x} & -5xe^{-5x} + e^{-5x} \end{vmatrix} = e^{-10x}.$$

Identifying $f(x) = e^{-10x}/x^2$ we obtain

$$u_1' = -\frac{xe^{-5x}e^{-10x}}{x^2 e^{-10x}} = -\frac{e^{-5x}}{x}$$

$$u_2' = \frac{e^{-5x}e^{-10x}}{x^2 e^{-10x}} = \frac{e^{-5x}}{x^2}.$$

Then

$$u_1 = -\int_{x_0}^{x} \frac{e^{-5t}}{t}\, dt, \quad x_0 > 0$$

$$u_2 = \int_{x_0}^{x} \frac{e^{-5t}}{t^2}\, dt, \quad x_0 > 0$$

and

$$y = c_1 e^{-5x} + c_2 x e^{-5x} - e^{-5x} \int_{x_0}^{x} \frac{e^{-5t}}{t}\, dt + xe^{-5x} \int_{x_0}^{x} \frac{e^{-5t}}{t^2}\, dt$$

for $x > 0$.

21. The auxiliary equation is $m^3 + m = m(m^2 + 1) = 0$, so $y_c = c_1 + c_2 \cos x + c_3 \sin x$ and

$$W = \begin{vmatrix} 1 & \cos x & \sin x \\ 0 & -\sin x & \cos x \\ 0 & -\cos x & -\sin x \end{vmatrix} = 1.$$

Identifying $f(x) = \tan x$ we obtain

$$u_1' = W_1 = \begin{vmatrix} 0 & \cos x & \sin x \\ 0 & -\sin x & \cos x \\ \tan x & -\cos x & -\sin x \end{vmatrix} = \tan x$$

$$u_2' = W_2 = \begin{vmatrix} 1 & 0 & \sin x \\ 0 & 0 & \cos x \\ 0 & \tan x & -\sin x \end{vmatrix} = -\sin x$$

$$u_3' = W_3 = \begin{vmatrix} 1 & \cos x & 0 \\ 0 & -\sin x & 0 \\ 0 & -\cos x & \tan x \end{vmatrix} = -\sin x \tan x = \frac{\cos^2 x - 1}{\cos x} = \cos x - \sec x.$$

Then

$$u_1 = -\ln|\cos x|$$

$$u_2 = \cos x$$

$$u_3 = \sin x - \ln|\sec x + \tan x|$$

and

$$y = c_1 + c_2 \cos x + c_3 \sin x - \ln|\cos x| + \cos^2 x$$

$$+ \sin^2 x - \sin x \ln|\sec x + \tan x|$$

$$= c_4 + c_2 \cos x + c_3 \sin x - \ln|\cos x| - \sin x \ln|\sec x + \tan x|$$

for $-\infty < x < \infty$.

24. The auxiliary equation is $2m^3 - 6m^2 = 2m^2(m - 3) = 0$, so $y_c = c_1 + c_2 x + c_3 e^{3x}$ and

$$W = \begin{vmatrix} 1 & x & e^{3x} \\ 0 & 1 & 3e^{3x} \\ 0 & 0 & 9e^{3x} \end{vmatrix} = 9e^{3x}.$$

Identifying $f(x) = x^2/2$ we obtain

$$u_1' = \frac{1}{9e^{3x}} W_1 = \frac{1}{9e^{3x}} \begin{vmatrix} 0 & x & e^{3x} \\ 0 & 1 & 3e^{3x} \\ x^2/2 & 0 & 9e^{3x} \end{vmatrix} = \frac{\frac{3}{2}x^3 e^{3x} - \frac{1}{2}x^2 e^{3x}}{9e^{3x}} = \frac{1}{6}x^3 - \frac{1}{18}x^2$$

$$u_2' = \frac{1}{9e^{3x}} W_2 = \frac{1}{9e^{3x}} \begin{vmatrix} 1 & 0 & e^{3x} \\ 0 & 0 & 3e^{3x} \\ 0 & x^2/2 & 9e^{3x} \end{vmatrix} = \frac{-\frac{3}{2}x^2 e^{3x}}{9e^{3x}} = -\frac{1}{6}x^2$$

$$u_3' = \frac{1}{9e^{3x}} W_3 = \frac{1}{9e^{3x}} \begin{vmatrix} 1 & x & 0 \\ 0 & 1 & 0 \\ 0 & 0 & x^2/2 \end{vmatrix} = \frac{\frac{1}{2}x^2}{9e^{3x}} = \frac{1}{18}x^2 e^{-3x}.$$

Then

$$u_1 = \frac{1}{24}x^4 - \frac{1}{54}x^3$$

$$u_2 = -\frac{1}{18}x^3$$

$$u_3 = -\frac{1}{54}x^2 e^{-3x} - \frac{1}{81}xe^{-3x} - \frac{1}{243}e^{-3x}$$

and

$$y = c_1 + c_2 x + c_3 e^{3x} + \frac{1}{24}x^4 - \frac{1}{54}x^3 - \frac{1}{18}x^4 - \frac{1}{54}x^2 - \frac{1}{81}x - \frac{1}{243}$$

$$= c_4 + c_5 x + c_3 e^{3x} - \frac{1}{72}x^4 - \frac{1}{54}x^3 - \frac{1}{54}x^2$$

for $-\infty < x < \infty$.

27. The auxiliary equation is $m^2 + 2m - 8 = (m-2)(m+4) = 0$, so $y_c = c_1 e^{2x} + c_2 e^{-4x}$ and

$$W = \begin{vmatrix} e^{2x} & e^{-4x} \\ 2e^{2x} & -4e^{-4x} \end{vmatrix} = -6e^{-2x}.$$

Identifying $f(x) = 2e^{-2x} - e^{-x}$ we obtain

$$u_1' = \frac{1}{3}e^{-4x} - \frac{1}{6}e^{-3x}$$

$$u_2' = -\frac{1}{6}e^{3x} - \frac{1}{3}e^{2x}.$$

Then

$$u_1 = -\frac{1}{12}e^{-4x} + \frac{1}{18}e^{-3x}$$

$$u_2 = \frac{1}{18}e^{3x} - \frac{1}{6}e^{2x}.$$

Thus

$$y = c_1 e^{2x} + c_2 e^{-4x} - \frac{1}{12} e^{-2x} + \frac{1}{18} e^{-x} + \frac{1}{18} e^{-x} - \frac{1}{6} e^{-2x}$$

$$= c_1 e^{2x} + c_2 e^{-4x} - \frac{1}{4} e^{-2x} + \frac{1}{9} e^{-x}$$

and

$$y' = 2c_1 e^{2x} - 4c_2 e^{-4x} + \frac{1}{2} e^{-2x} - \frac{1}{9} e^{-x}.$$

The initial conditions imply

$$c_1 + c_2 - \frac{5}{36} = 1$$

$$2c_1 - 4c_2 + \frac{7}{18} = 0.$$

Thus $c_1 = 25/36$ and $c_2 = 4/9$, and

$$y = \frac{25}{36} e^{2x} + \frac{4}{9} e^{-4x} - \frac{1}{4} e^{-2x} + \frac{1}{9} e^{-x}.$$

30. Write the equation in the form

$$y'' - \frac{4}{x} y' + \frac{6}{x^2} y = \frac{1}{x^3}$$

and identify $f(x) = 1/x^3$. From $y_1 = x^2$ and $y_2 = x^3$ we compute

$$W(y_1, y_2) = \begin{vmatrix} x^2 & x^3 \\ 2x & 3x^2 \end{vmatrix} = 3x^4 - 2x^4 = x^4.$$

Now

$$u_1' = -\frac{x^3/x^3}{x^4} = -\frac{1}{x^4} \quad \text{so} \quad u_1 = \frac{1}{3x^3},$$

and

$$u_2' = \frac{x^2/x^3}{x^4} = \frac{1}{x^5} \quad \text{so} \quad u_2 = \frac{1}{4x^4}.$$

Thus

$$y_p = \frac{x^2}{3x^3} - \frac{x^3}{4x^4} = \frac{1}{12x}$$

and

$$y = y_c + y_p = c_1 x^2 + c_2 x^3 + \frac{1}{12x}.$$

33. (a) We have $y_c = c_1 e^{-x} + c_2 x e^{-x}$ and we assume $y_p = Ax^2 + Bx + C$. Substituting into the differential equation we find

$$A = 4$$

$$4A + B = 0$$

$$2A + 2B + C = -3$$

so that $A = 4$, $B = -16$, and $C = 21$. A particular solutionis $y_p = 4x^2 - 16x + 21$.

(b) We have $y_c = c_1 e^{-x} + c_2 x e^{-x}$ and

$$W = \begin{vmatrix} e^{-x} & xe^{-x} \\ -e^{-x} & -xe^{-x} + e^{-x} \end{vmatrix} = e^{-2x}.$$

Identifying $f(x) = e^{-x}/x$ we obtain

$$u_1' = -\frac{xe^{-x}e^{-x}/x}{e^{-2x}} = -1$$

$$u_2' = \frac{e^{-x}e^{-x}/x}{e^{-2x}} = \frac{1}{x}.$$

Then $u_1 = -x$, $u_2 = \ln x$ and

$$y_p = -xe^{-x} + xe^{-x}\ln x.$$

Since $-xe^{-x}$ is a solution of the homogeneous differential equation, we take $y_p = xe^{-x}\ln x$.

(c) Adding the results of **(a)** and **(b)** we have

$$y_p = 4x^2 - 16x + 21 + xe^{-x}\ln x.$$

Chapter 4 Review Exercises

3. False; consider $f_1(x) = 0$ and $f_2(x) = x$. These are linearly dependent even though x is not a multiple of 0. The statement would be true if it read: "Two functions $f_1(x)$ and $f_2(x)$ are linearly independent on an interval if *neither* is a constant multiple of the other."

6. True

9. $A + Bxe^x$

12. Identifying $P(x) = -2 - 2/x$ we have $\int P\, dx = -2x - 2\ln x$ and

$$y_2 = e^x \int \frac{e^{2x + \ln x^2}}{e^{2x}}\, dx = e^x \int x^2\, dx = \frac{1}{3}x^3 e^x.$$

15. From $m^3 + 10m^2 + 25m = 0$ we obtain $m = 0$, $m = -5$, and $m = -5$ so that

$$y = c_1 + c_2 e^{-5x} + c_3 x e^{-5x}.$$

18. From $2m^4 + 3m^3 + 2m^2 + 6m - 4 = 0$ we obtain $m = 1/2$, $m = -2$, and $m = \pm\sqrt{2}\,i$ so that

$$y = c_1 e^{x/2} + c_2 e^{-2x} + c_3 \cos\sqrt{2}\,x + c_4 \sin\sqrt{2}\,x.$$

21. Applying $D(D^2 + 1)$ to the differential equation we obtain

$$D(D^2 + 1)(D^3 - 5D^2 + 6D) = D^2(D^2 + 1)(D - 2)(D - 3) = 0.$$

Then

$$y = \underbrace{c_1 + c_2 e^{2x} + c_3 e^{3x}}_{y_c} + c_4 x + c_5 \cos x + c_6 \sin x$$

and $y_p = Ax + B\cos x + C\sin x$. Substituting y_p into the differential equation yields

$$6A + (5B + 5C)\cos x + (-5B + 5C)\sin x = 8 + 2\sin x.$$

Equating coefficients gives $A = 4/3$, $B = -1/5$, and $C = 1/5$. The general solution is

$$y = c_1 + c_2 e^{2x} + c_3 e^{3x} + \frac{4}{3}x - \frac{1}{5}\cos x + \frac{1}{5}\sin x.$$

24. The auxiliary equation is $m^2 - 1 = (m-1)(m+1) = 0$ so that $m = \pm 1$ and $y = c_1 e^x + c_2 e^{-x}$. Assuming $y_p = Ax + B + C\sin x$ and substituting into the differential equation we find $A = -1$, $B = 0$, and $C = -\frac{1}{2}$. Thus $y_p = -x - \frac{1}{2}\sin x$ and

$$y = c_1 e^x + c_2 e^{-x} - x - \frac{1}{2}\sin x.$$

Setting $y(0) = 2$ and $y'(0) = 3$ we obtain

$$c_1 + c_2 = 2$$

$$c_1 - c_2 - \frac{3}{2} = 3.$$

Solving this system we find $c_1 = \frac{13}{4}$ and $c_2 = -\frac{5}{4}$. The solution of the initial-value problem is

$$y = \frac{13}{4}e^x - \frac{5}{4}e^{-x} - x - \frac{1}{2}\sin x.$$

27. The auxiliary equation is $2m^3 - 13m^2 + 24m - 9 = (2m - 1)(m - 3)^2 = 0$ so that

$$y_c = c_1 e^{x/2} + c_2 e^{3x} + c_3 x e^{3x}.$$

A particular solution is $y_p = -4$ and the general solution is

$$y = c_1 e^{x/2} + c_2 e^{3x} + c_3 x e^{3x} - 4.$$

Setting $y(0) = -4$, $y'(0) = 0$, and $y''(0) = \frac{5}{2}$ we obtain

$$c_1 + c_2 - 4 = -4$$

$$\frac{1}{2}c_1 + 3c_2 + c_3 = 0$$

$$\frac{1}{4}c_1 + 9c_2 + 6c_3 = \frac{5}{2}.$$

Solving this system we find $c_1 = \frac{2}{5}$, $c_2 = -\frac{2}{5}$, and $c_3 = 1$. Thus

$$y = \frac{2}{5}e^{x/2} - \frac{2}{5}e^{3x} + xe^{3x} - 4.$$

5 Applications of Second-Order Differential Equations: Vibrational Models

_____ **Exercises 5.1** _____

3. Applying the initial conditions to $x(t) = c_1 \cos 5t + c_2 \sin 5t$ and $x'(t) = -5c_1 \sin 5t + 5c_2 \cos 5t$ gives

$$x(0) = c_1 = -2 \quad \text{and} \quad x'(0) = 5c_2 = 10.$$

Then $c_1 = -2$, $c_2 = 2$, and

$$A = \sqrt{4+4} = 2\sqrt{2} \quad \text{and} \quad \tan\phi = \frac{-2}{2} = -1.$$

Since $\sin\phi < 0$ and $\cos\phi > 0$, ϕ is a fourth quadrant angle and $\phi = -\pi/4$. Thus $x(t) = 2\sqrt{2}\sin(5t - \pi/4)$.

6. Applying the initial conditions to $x(t) = c_1 \cos 8t + c_2 \sin 8t$ and $x'(t) = -8c_1 \sin 8t + 8c_2 \cos 8t$ gives

$$x(0) = c_1 = 4 \quad \text{and} \quad x'(0) = 8c_2 = 16.$$

Then $c_1 = 4$, $c_2 = 2$, and

$$A = \sqrt{16+4} = 2\sqrt{5} \quad \text{and} \quad \tan\phi = \frac{4}{2} = 2.$$

Since $\sin\phi > 0$ and $\cos\phi > 0$, ϕ is a first quadrant angle and $\phi = \tan^{-1}2 \approx 1.107$. Thus $x(t) \approx 2\sqrt{5}\sin(8t + 1.107)$.

9. From $mx'' + 16x = 0$ we obtain

$$x = c_1 \cos\frac{4}{\sqrt{m}}t + c_2 \sin\frac{4}{\sqrt{m}}t$$

so that the period $\pi/4 = \pi\sqrt{m}/2$, $m = 1/4$ slug, and the weight is 8 lb.

12. From $20x'' + kx = 0$ we obtain

$$x = c_1 \cos\frac{1}{2}\sqrt{\frac{k}{5}}t + c_2 \sin\frac{1}{2}\sqrt{\frac{k}{5}}t$$

so that the frequency $2/\pi = \frac{1}{4}\sqrt{k/5}\,\pi$ and $k = 320$ N/m. If $80x'' + 320x = 0$ then $x = c_1 \cos 2t + c_2 \sin 2t$ so that the frequency is $2/2\pi = 1/\pi$ vibrations/second.

15. From $\frac{5}{8}x'' + 40x = 0$, $x(0) = 1/2$, and $x'(0) = 0$ we obtain $x = \frac{1}{2}\cos 8t$.

(a) $x(\pi/12) = -1/4$, $x(\pi/8) = -1/2$, $x(\pi/6) = -1/4$, $x(\pi/8) = 1/2$, $x(9\pi/32) = \sqrt{2}/4$.

(b) $x' = -4\sin 8t$ so that $x'(3\pi/16) = 4$ ft/s directed downward.

(c) If $x = \frac{1}{2}\cos 8t = 0$ then $t = (2n+1)\pi/16$ for $n = 0, 1, 2, \ldots$.

18. From $x'' + 16x = 0$, $x(0) = -1$, and $x'(0) = -2$ we obtain

$$x = -\cos 4t - \frac{1}{2}\sin 4t = \frac{\sqrt{5}}{2}\cos(4t - 3.6).$$

The period is $\pi/2$ seconds and the amplitude is $\sqrt{5}/2$ feet. In 4π seconds it will make 8 complete vibrations.

21. From $2x'' + 200x = 0$, $x(0) = -2/3$, and $x'(0) = 5$ we obtain

(a) $x = -\frac{2}{3}\cos 10t + \frac{1}{2}\sin 10t = \frac{5}{6}\sin(10t - 0.927)$.

(b) The amplitude is $5/6$ ft and the period is $2\pi/10 = \pi/5$

(c) $3\pi = \pi k/5$ and $k = 15$ cycles.

(d) If $x = 0$ and the weight is moving downward for the second time, then $10t - 0.927 = 2\pi$ or $t = 0.721$ s.

(e) If $x' = \frac{25}{3}\cos(10t - 0.927) = 0$ then $10t - 0.927 = \pi/2 + n\pi$ or $t = (2n+1)\pi/20 + 0.0927$ for $n = 0, 1, 2, \ldots$.

(f) $x(3) = -0.597$ ft

(g) $x'(3) = -5.814$ ft/s

(h) $x''(3) = 59.702$ ft/s^2

(i) If $x = 0$ then $t = \frac{1}{10}(0.927 + n\pi)$ for $n = 0, 1, 2, \ldots$ and $x'(t) = \pm\frac{25}{3}$ ft/s.

(j) If $x = 5/12$ then $t = \frac{1}{10}(\pi/6 + 0.927 + 2n\pi)$ and $t = \frac{1}{10}(5\pi/6 + 0.927 + 2n\pi)$ for $n = 0, 1, 2, \ldots$.

(k) If $x = 5/12$ and $x' < 0$ then $t = \frac{1}{10}(5\pi/6 + 0.927 + 2n\pi)$ for $n = 0, 1, 2, \ldots$.

24. Let m denote the mass in slugs of the first weight. Let k_1 and k_2 be the spring constants and $k = 4k_1k_2/(k_1 + k_2)$ the effective spring constant of the system. Now, the numerical value of the first weight is $W = mg = 32m$, so

$$32m = k_1\left(\frac{1}{3}\right) \quad \text{and} \quad 32m = k_2\left(\frac{1}{2}\right).$$

From these equations we find $2k_1 = 3k_2$. The given period of the combined system is $2\pi/w = \pi/15$, so $w = 30$. Since the mas of an 8-pound weight is $1/4$ slug, we have from $w^2 = k/m$

$$30^2 = \frac{k}{1/4} = 4k \quad \text{or} \quad k = 225.$$

We now have the system of equations

$$\frac{4k_1k_2}{k_1 + k_2} = 225$$

$$2k_1 = 3k_2.$$

Solving the second equation for k_1 and substituting in the first equation, we obtain

$$\frac{4(3k_2/2)k_2}{3k_2/2 + k_2} = \frac{12k_2^2}{5k_2} = \frac{12k_2}{5} = 225.$$

Thus, $k_2 = 375/4$ and $k_1 = 1125/8$. Finally, the value of the first weight is

$$W = 32m = \frac{k_1}{3} = \frac{1125/8}{3} = \frac{375}{8} \approx 46.88 \text{ lb.}$$

27. $x = 2\sqrt{2}\left[\frac{-1}{\sqrt{2}}\cos 5t - \frac{-1}{\sqrt{2}}\sin 5t\right] = 2\sqrt{2}\cos(5t + 5\pi/4)$.

30. If $x = A\sin(\omega t + \phi)$ the extremes for x occur when $x' = A\omega\cos(\omega t + \phi) = 0$, or $t = (\pi/2 - \phi + 2n\pi)\frac{1}{\omega}$ and $t = (-\pi/2 - \phi + 2n\pi)$ for $n = 0, 1, 2, \ldots$. Thus, the time interval between successive maxima is $2\pi/\omega$.

Exercises 5.2

3. (a) above   **(b)** heading upward

6. (a) above   **(b)** heading downward

9. (a) From $x'' + 10x' + 16x = 0$, $x(0) = 1$, and $x'(0) = 0$ we obtain $x = \frac{4}{3}e^{-2t} - \frac{1}{3}e^{-8t}$.

     (b) From $x'' + x' + 16x = 0$, $x(0) = 1$, and $x'(0) = -12$ then $x = -\frac{2}{3}e^{-2t} + \frac{5}{3}e^{-8t}$.

12. (a) From $\frac{1}{4}x'' + x' + 5x = 0$, $x(0) = 1/2$, and $x'(0) = 1$ we obtain $x = e^{-2t}\left(\frac{1}{2}\cos 4t + \frac{1}{2}\sin 4t\right)$.

     (b) $x = e^{-2t}\frac{1}{\sqrt{2}}\left(\frac{\sqrt{2}}{2}\cos 4t + \frac{\sqrt{2}}{2}\sin 4t\right) = \frac{1}{\sqrt{2}}e^{-2t}\sin\left(4t + \frac{\pi}{4}\right)$.

     (c) If $x = 0$ then $4t + \pi/4 = \pi, 2\pi, 3\pi, \ldots$ so that the times heading downward are $t = (7 + 8n)\pi/16$ for $n = 0, 1, 2, \ldots$.

     (d)

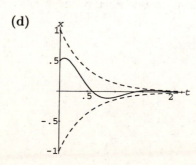

15. From $40x'' + 560x' + 3920x = 0$, $x(0) = 0$, and $x'(0) = 2$ we obtain

$$x = \frac{2}{7}e^{-7t}\sin 7t.$$

18. From $x'' + \beta x' + 25x = 0$ we see that the roots of the auxiliary equation are $m = -\frac{\beta}{2} \pm \frac{1}{2}\sqrt{100 - \beta^2}\, i$. The quasi-period is $\pi/2 = 2\pi \left/ \frac{1}{2}\sqrt{100 - \beta^2}\right.$ so that $\beta = 6$.

21. The time interval between successive values of t for which (15) touches the graphs of $y = \pm Ae^{-\lambda t}$ is

$$t = \frac{(2n+3)\pi/2 - \phi}{\sqrt{\omega^2 - \lambda^2}} - \frac{(2n+1)\pi/2 - \phi}{\sqrt{\omega^2 - \lambda^2}} = \frac{\pi}{\sqrt{\omega^2 - \lambda^2}}.$$

24. (a) If $\delta > 0$ is very small then x_n is slightly larger than x_{n+2} and the rate of damping is slow.

(b) If $x = \frac{1}{\sqrt{2}}e^{-2t}\sin(4t + \pi/4)$ then $\delta = 2\pi\lambda/\sqrt{\omega^2 - \lambda^2} = 4\pi/4 = \pi$.

Exercises 5.3

3. From $x'' + 8x' + 16x = 8\sin 4t$, $x(0) = 0$, and $x'(0) = 0$ we obtain $x_c = c_1 e^{-4t} + c_2 t e^{-4t}$ and $x_p = -\frac{1}{4}\cos 4t$ so that the equation of motion is

$$x = \frac{1}{4}e^{-4t} + te^{-4t} - \frac{1}{4}\cos 4t.$$

6. Since $x = \frac{\sqrt{85}}{4}\sin(4t - 0.219) - \frac{\sqrt{17}}{2}e^{-2t}\sin(4t - 2.897)$, the amplitude approaches $\sqrt{85}/4$ as $t \to \infty$.

9. (a) From $100x'' + 1600x = 1600\sin 8t$, $x(0) = 0$, and $x'(0) = 0$ we obtain $x_c = c_1\cos 4t + c_2\sin 4t$ and $x_p = -\frac{1}{3}\sin 8t$ so that

$$x = \frac{2}{3}\sin 4t - \frac{1}{3}\sin 8t.$$

(b) If $x = \frac{1}{3}\sin 4t(2 - 2\cos 4t) = 0$ then $t = n\pi/4$ for $n = 0, 1, 2, \ldots$.

(c) If $x' = \frac{8}{3}\cos 4t - \frac{8}{3}\cos 8t = \frac{8}{3}(1 - \cos 4t)(1 + 2\cos 4t) = 0$ then $t = \pi/3 + n\pi/2$ and $t = \pi/6 + n\pi/2$ for $n = 0, 1, 2, \ldots$ at the extreme values. *Note:* There are many other values of t for which $x' = 0$.

(d) $x(\pi/6 + n\pi/2) = \sqrt{3}/2$ cm. and $x(\pi/3 + n\pi/2) = -\sqrt{3}/2$ cm.

(e)

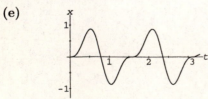

12. (a) If $x'' + \beta x' + 3x = 0$ and $0 < \beta < 2\sqrt{3}$ then the roots of the auxiliary equation are

$$m = \tfrac{1}{2}\left(-\beta \pm \sqrt{\beta^2 - 12}\right);$$ this is underdamped motion. The system is in resonance when

$\gamma = \sqrt{3 - \beta^2/2}$, where we require that $3 - \beta/2 > 0$, or $0 < \beta < \sqrt{6}$.

(b) When $F_0 = 3$, the resonance curve is given by

$$g(\gamma) = \frac{3}{\sqrt{(3 - \gamma^2)^2 + \beta^2 \gamma^2}},$$

and the family of graphs is shown for various values of β.

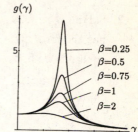

15. (a) From $x'' + \omega^2 x = F_0 \cos \gamma t$, $x(0) = 0$, and $x'(0) = 0$ we obtain $x_c = c_1 \cos \omega t + c_2 \sin \omega t$ and $x_p = (F_0 \cos \gamma t)/(\omega^2 - \gamma^2)$ so that

$$x = -\frac{F_0}{\omega^2 - \gamma^2} \cos \omega t + \frac{F_0}{\omega^2 - \gamma^2} \cos \gamma t.$$

(b) $\displaystyle \lim_{\gamma \to \omega} \frac{F_0}{\omega^2 - \gamma^2}(\cos \gamma t - \cos \omega t) = \lim_{\gamma \to \omega} \frac{-F_0 t \sin \gamma t}{-2\gamma} = \frac{F_0}{2\omega} t \sin \omega t.$

18. From $x'' + 9x = 5 \sin 3t$, $x(0) = 2$, and $x'(0) = 0$ we obtain $x_c = c_1 \cos 3t + c_2 \sin 3t$, $x_p = -\tfrac{5}{6}t \cos 3t$, and

$$x = 2 \cos 3t + \frac{5}{18} \sin 3t - \frac{5}{6}t \cos 3t.$$

_____ Exercises 5.4 _____

3. Since $R^2 - 4L/C = -20 < 0$, the circuit is underdamped.

6. Solving $\tfrac{1}{4}q'' + 20q' + 300q = 0$ we obtain $q(t) = c_1 e^{-20t} + c_2 e^{-60t}$. The initial conditions $q(0) = 4$ and $q'(0) = 0$ imply $c_1 = 6$ and $c_2 = -2$. Thus

$$q(t) = 6e^{-20t} - 2e^{-60t}.$$

Setting $q = 0$ we find $e^{40t} = 1/3$ which implies $t < 0$. Therefore the charge is never 0.

9. Solving $q'' + 2q' + 4q = 0$ we obtain $y_c = e^{-t}\left(\cos \sqrt{3}\, t + \sin \sqrt{3}\, t\right)$. The steady-state charge has the form $y_p = A \cos t + B \sin t$. Substituting into the differential equation we find

$$(3A + 2B)\cos t + (3B - 2A)\sin t = 50 \cos t.$$

Thus, $A = 150/13$ and $B = 100/13$. The steady-state charge is

$$q_p(t) = \frac{150}{13} \cos t + \frac{100}{13} \sin t.$$

and the steady-state current is

$$i_p(t) = -\frac{150}{13}\sin t + \frac{100}{13}\cos t.$$

12. Solving $\frac{1}{2}q'' + 20q' + 1000q = 0$ we obtain $q_c(t) = (c_1\cos 40t + c_2\sin 40t)$. The steady-state charge has the form $q_p(t) = A\sin 60t + B\cos 60t + C\sin 40t + D\cos 40t$. Substituting into the differential equation we find

$$(-1600A - 2400B)\sin 60t + (2400A - 1600B)\cos 60t$$

$$+ (400C - 1600D)\sin 40t + (1600C + 400D)\cos 40t$$

$$= 200\sin 60t + 400\cos 40t.$$

Equating coefficients we obtain $A = -1/26$, $B = -3/52$, $C = 4/17$, and $D = 1/17$. The steady-state charge is

$$q_p(t) = -\frac{1}{26}\sin 60t - \frac{3}{52}\cos 60t + \frac{4}{17}\sin 40t + \frac{1}{17}\cos 40t$$

and the steady-state current is

$$i_p(t) = -\frac{30}{13}\cos 60t + \frac{45}{13}\sin 60t + \frac{160}{17}\cos 40t - \frac{40}{17}\sin 40t.$$

15. By Problem 10 the amplitude of the steady-state current is E_0/Z, where $Z = \sqrt{X^2 + R^2}$ and $X = L\gamma - 1/C\gamma$. Since E_0 is constant the amplitude will be a maximum when Z is a minimum. Since R is constant, Z will be a minimum when $X = 0$. Solving $L\gamma - 1/C\gamma = 0$ for C we obtain $C = 1/L\gamma^2$.

18. When the circuit is in resonance the form of $q_p(t)$ is $q_p(t) = At\cos kt + Bt\sin kt$ where $k = 1/\sqrt{LC}$. Substituting $q_p(t)$ into the differential equation we find

$$q_p'' + k^2q = -2kA\sin kt + 2kB\cos kt = \frac{E_0}{L}\cos kt.$$

Equating coefficients we obtain $A = 0$ and $B = E_0/2kL$. The charge is

$$q(t) = c_1\cos kt + c_2\sin kt + \frac{E_0}{2kL}t\sin kt.$$

The initial conditions $q(0) = q_0$ and $q'(0) = i_0$ imply $c_1 = q_0$ and $c_2 = i_0/k$. The current is

$$i(t) = -c_1k\sin kt + c_2k\cos kt + \frac{E_0}{2kL}(kt\cos kt + \sin kt)$$

$$= \left(\frac{E_0}{2kL} - q_0k\right)\sin kt + i_0\cos kt + \frac{E_0}{2L}t\cos kt.$$

───────── **Chapter 5 Review Exercises** ─────────

3. 5/4 m., since $x = -\cos 4t + \frac{3}{4}\sin 4t$.

6. False

9. 9/2, since $x = c_1 \cos \sqrt{2k}\, t + c_2 \sin \sqrt{2k}\, t$.

12. From $x'' + \beta x' + 64x = 0$ we see that oscillatory motion results if $\beta^2 - 256 < 0$ or $0 \le |\beta| < 16$.

15. Writing $\frac{1}{8}x'' + \frac{8}{3}x = \cos \gamma t + \sin \gamma t$ in the form $x'' + \frac{64}{3}x = 8\cos \gamma t + 8\sin \gamma t$ we identify $\lambda = 0$ and $\omega^2 = 64/3$. From Example 4 in Section 5.3 we see that the system is in a state of pure resonance when $\gamma = \sqrt{64/3} = 8/\sqrt{3}$.

18. (a) Let k be the effective spring constant and x_1 and x_2 the elongation of springs k_1 and k_2. The restoring forces satisfy $k_1 x_1 = k_2 x_2$ so $x_2 = (k_1/k_2)x_1$. From $k(x_1 + x_2) = k_1 x_1$ we have

$$k\left(x_1 + \frac{k_1}{k_2}x_2\right) = k_1 x_1$$

$$k\left(\frac{k_2 + k_1}{k_2}\right) = k_1$$

$$k = \frac{k_1 k_2}{k_1 + k_2}$$

$$\frac{1}{k} = \frac{1}{k_1} + \frac{1}{k_2}.$$

(b) From $k_1 = 2W$ and $k_2 = 4W$ we find $1/k = 1/2W + 1/4W = 3/4W$. Then $k = 4W/3 = 4mg/3$. The differential equation $mx'' + kx = 0$ then becomes $x'' + (4g/3)x = 0$. The solution is

$$x(t) = c_1 \cos 2\sqrt{\frac{g}{3}}\, t + c_2 \sin 2\sqrt{\frac{g}{3}}\, t.$$

The initial conditions $x(0) = 1$ and $x'(0) = 2/3$ imply $c_1 = 1$ and $c_2 = 1/\sqrt{3g}$.

(c) To compute the maximum speed of the weight we compute

$$x'(t) = 2\sqrt{\frac{g}{3}}\sin 2\sqrt{\frac{g}{3}}\, t + \frac{2}{3}\cos 2\sqrt{\frac{g}{3}}\, t \quad \text{and} \quad |x'(t)| = \sqrt{4\frac{g}{3} + \frac{4}{9}} = \frac{2}{3}\sqrt{3g + 1}.$$

6 Differential Equations with Variable Coefficients

_____ **Exercises 6.1** _____

3. The auxiliary equation is $m^2 = 0$ so that $y = c_1 + c_2 \ln x$.

6. The auxiliary equation is $m^2 + 4m + 3 = (m+1)(m+3) = 0$ so that $y = c_1 x^{-1} + c_2 x^{-3}$.

9. The auxiliary equation is $25m^2 + 1 = 0$ so that $y = c_1 \cos\left(\frac{1}{5}\ln x\right) + c_2 \left(\frac{1}{5}\ln x\right)$.

12. The auxiliary equation is $m^2 + 7m + 6 = (m+1)(m+6) = 0$ so that $y = c_1 x^{-1} + c_2 x^{-6}$.

15. The auxiliary equation is $3m^2 + 3m + 1 = 0$ so that $y = x^{-1/2}\left[c_1 \cos\left(\frac{\sqrt{3}}{6}\ln x\right) + c_2 \sin\left(\frac{\sqrt{3}}{6}\ln x\right)\right]$.

18. Assuming that $y = x^m$ and substituting into the differential equation we obtain

$$m(m-1)(m-2) + m - 1 = m^3 - 3m^2 + 3m - 1 = (m-1)^3 = 0.$$

Thus

$$y = c_1 x + c_2 x \ln x + c_3 x(\ln x)^2.$$

21. Assuming that $y = x^m$ and substituting into the differential equation we obtain

$$m(m-1)(m-2)(m-3) + 6m(m-1)(m-2) = m^4 - 7m^2 + 6m = m(m-1)(m-2)(m+3) = 0.$$

Thus

$$y = c_1 + c_2 x + c_3 x^2 + c_4 x^{-3}.$$

24. The auxiliary equation is $m^2 - 6m + 8 = (m-2)(m-4) = 0$, so that

$$y = c_1 x^2 + c_2 x^4 \quad \text{and} \quad y' = 2c_1 x + 4c_2 x^3.$$

The initial conditions imply

$$4c_1 + 16c_2 = 32$$

$$4c_1 + 32c_2 = 0.$$

Thus, $c_1 = 16$, $c_2 = -2$, and $y = 16x^2 - 2x^4$.

27. In this problem we use the substitution $t = -x$ since the initial conditions are on the interval $(-\infty, 0)$. Then

$$\frac{dy}{dt} = \frac{dy}{dx}\frac{dx}{dt} = -\frac{dy}{dx}$$

and

$$\frac{d^2y}{dt^2} = \frac{d}{dt}\left(\frac{dy}{dt}\right) = \frac{d}{dt}\left(-\frac{dy}{dx}\right) = -\frac{d}{dt}(y') = -\frac{dy'}{dx}\frac{dx}{dt} = -\frac{d^2y}{dx^2}\frac{dx}{dt} = \frac{d^2y}{dx^2}.$$

The differential equation and initial conditions become

$$4t^2\frac{d^2y}{dt^2} + y = 0; \quad y(t)\Big|_{t=1} = 2, \quad y'(t)\Big|_{t=1} = -4.$$

The auxiliary equation is $4m^2 - 4m + 1 = (2m-1)^2 = 0$, so that

$$y = c_1 t^{1/2} + c_2 t^{1/2}\ln t \quad \text{and} \quad y' = \frac{1}{2}c_1 t^{-1/2} + c_2\left(t^{-1/2} + \frac{1}{2}t^{-1/2}\ln t\right).$$

The initial conditions imply $c_1 = 2$ and $1 + c_2 = -4$. Thus

$$y = 2t^{1/2} - 5t^{1/2}\ln t = 2(-x)^{1/2} - 5(-x)^{1/2}\ln(-x), \quad x < 0.$$

30. The auxiliary equation is $m^2 - 5m = m(m-5) = 0$ so that $y_c = c_1 + c_2 x^5$ and

$$W(1, x^5) = \begin{vmatrix} 1 & x^5 \\ 0 & 5x^4 \end{vmatrix} = 5x^4.$$

Identifying $f(x) = x^3$ we obtain $u_1' = -\frac{1}{5}x^4$ and $u_2' = 1/5x$. Then $u_1 = -\frac{1}{25}x^5$, $u_2 = \frac{1}{5}\ln x$, and

$$y = c_1 + c_2 x^5 - \frac{1}{25}x^5 + \frac{1}{5}x^5\ln x = c_1 + c_3 x^5 + \frac{1}{5}x^5\ln x.$$

33. The auxiliary equation is $m^2 - 2m + 1 = (m-1)^2 = 0$ so that $y_c = c_1 x + c_2 x\ln x$ and

$$W(x, x\ln x) = \begin{vmatrix} x & x\ln x \\ 1 & 1 + \ln x \end{vmatrix} = x.$$

Identifying $f(x) = 2/x$ we obtain $u_1' = -2\ln x/x$ and $u_2' = 2/x$. Then $u_1 = -(\ln x)^2$, $u_2 = 2\ln x$, and

$$y = c_1 x + c_2 x\ln x - x(\ln x)^2 + 2x(\ln x)^2 = c_1 x + c_2 x\ln x + x(\ln x)^2.$$

36. From Example 6 in the text: When $x = e^t$ or $t = \ln x$,

$$\frac{dy}{dx} = \frac{1}{x}\frac{dy}{dt} \quad \text{and} \quad \frac{d^2y}{dx^2} = \frac{1}{x^2}\left[\frac{d^2y}{dt^2} - \frac{dy}{dt}\right].$$

Substituting into the differential equation we obtain

$$\frac{d^2y}{dt^2} - 5\frac{dy}{dt} + 6y = 2t.$$

The auxiliary equation is $m^2 - 5m + 6 = (m-2)(m-3) = 0$ so that $y_c = c_1 e^{2t} + c_2 e^{3t}$. Using undetermined coefficients we try $y_p = At + B$. This leads to $(-5A + 6B) + 6At = 2t$, so that $A = 1/3$, $B = 5/18$, and

$$y = c_1 e^{2t} + c_2 e^{3t} + \frac{1}{3}t + \frac{5}{18} = c_1 x^2 + c_2 x^3 + \frac{1}{3}\ln x + \frac{5}{18}.$$

59

39. From Example 6 in the text: When $x = e^t$ or $t = \ln x$,

$$\frac{dy}{dx} = \frac{1}{x}\frac{dy}{dt} \quad \text{and} \quad \frac{d^2y}{dx^2} = \frac{1}{x^2}\left[\frac{d^2y}{dt^2} - \frac{dy}{dt}\right].$$

Substituting into the differential equation we obtain

$$\frac{d^2y}{dt^2} + 8\frac{dy}{dt} - 20y = 5e^{-3t}.$$

The auxiliary equation is $m^2 + 8m - 20 = (m+10)(m-2) = 0$ so that $y_c = c_1 e^{-10t} + c_2 e^{2t}$. Using undetermined coefficients we try $y_p = Ae^{-3t}$. This leads to $-35Ae^{-3t} = 5e^{-3t}$, so that $A = -1/7$ and

$$y = c_1 e^{-10t} + c_2 e^{2t} - \frac{1}{7}e^{-3t} = c_1 x^{-10} + c_2 x^2 - \frac{1}{7}x^{-3}.$$

42. The auxiliary equation is $m^2 = 0$ so that $u(r) = c_1 + c_2 \ln r$. The boundary conditions $u(a) = u_0$ and $u(b) = u_1$ yield the system $c_1 + c_2 \ln a = u_0$, $c_1 + c_2 \ln b = u_1$. Solving gives

$$c_1 = \frac{u_1 \ln a - u_0 \ln b}{\ln(a/b)} \quad \text{and} \quad c_2 = \frac{u_0 - u_1}{\ln(a/b)}.$$

Thus,

$$u(r) = \frac{u_1 \ln a - u_0 \ln b}{\ln(a/b)} + \frac{u_0 - u_1}{\ln(a/b)}\ln r = \frac{u_0 \ln(r/b) - u_1 \ln(r/a)}{\ln(a/b)}.$$

45. Letting $t = x + 2$ we obtain

$$\frac{dy}{dx} = \frac{dy}{dt}$$

and

$$\frac{d^2y}{dx^2} = \frac{d}{dx}\left(\frac{dy}{dt}\right) = \frac{d^2y}{dt^2}\frac{dt}{dx} = \frac{d^2y}{dt^2}.$$

Substituting into the differential equation we obtain

$$t^2\frac{d^2y}{dt^2} + t\frac{dy}{dt} + y = 0.$$

The auxiliary equation is $m^2 + 1 = 0$ so that

$$y = c_1 \cos(\ln t) + c_2 \sin(\ln t) = c_1 \cos\left[\ln(x+2)\right] + c_2 \sin\left[\ln(x+2)\right].$$

Exercises 6.2

3. $\lim\limits_{n\to\infty}\left|\dfrac{a_{n+1}}{a_n}\right| = \lim\limits_{n\to\infty}\left|\dfrac{2^{n+1}x^{n+1}/(n+1)}{2^n x^n/n}\right| = \lim\limits_{n\to\infty}\dfrac{2n}{n+1}|x| = 2|x|$

The series is absolutely convergent for $2|x| < 1$ or $|x| < 1/2$. At $x = -1/2$, the series $\sum\limits_{k=1}^{\infty}\dfrac{(-1)^k}{k}$ converges by the alternating series test. At $x = 1/2$, the series $\sum\limits_{k=1}^{\infty}\dfrac{1}{k}$ is the harmonic series which diverges. Thus, the given series converges on $[-1/2, 1/2)$.

6. $\lim\limits_{n\to\infty}\left|\dfrac{a_{n+1}}{a_n}\right| = \lim\limits_{n\to\infty}\left|\dfrac{(x+7)^{n+1}/\sqrt{n+1}}{(x+7)^n\sqrt{n}}\right| = \lim\limits_{n\to\infty}\sqrt{\dfrac{n}{n+1}}\,|x+7| = |x+7|$

The series is absolutely convergent for $|x+7| < 1$ or on $(-8, 6)$. At $x = -8$, the series $\sum\limits_{n=1}^{\infty}\dfrac{(-1)^n}{\sqrt{n}}$ converges by the alternating series test. At $x = -6$, the series $\sum\limits_{n=1}^{\infty}\dfrac{1}{\sqrt{n}}$ is a divergent p-series. Thus, the given series converges on $[-8, -6)$.

9. $\lim\limits_{n\to\infty}\left|\dfrac{a_{n+1}}{a_n}\right| = \lim\limits_{n\to\infty}\left|\dfrac{(n+1)!2^{n+1}x^{n+1}}{n!2^n x^n}\right| = \lim\limits_{n\to\infty}2(n+1)|x| = \infty,\ x\ne 0$

The series converges only at $x = 0$.

12. $e^{-x}\cos x = \left(1 - x + \dfrac{x^2}{2} - \dfrac{x^3}{6} + \dfrac{x^4}{24} - \cdots\right)\left(1 - \dfrac{x^2}{2} + \dfrac{x^4}{24} - \cdots\right) = 1 - x + \dfrac{x^3}{3} - \dfrac{x^4}{6} + \cdots$

15. $\left(x - \dfrac{x^3}{3} + \dfrac{x^5}{5} - \dfrac{x^7}{7} + \cdots\right)^2 = x^2 - \dfrac{2x^4}{3} + \dfrac{23x^6}{45} - \dfrac{44x^8}{105} + \cdots$

18. $e^x + e^{-x} = \left(1 + x + \dfrac{x^2}{2} + \dfrac{x^3}{6} + \dfrac{x^4}{24} + \dfrac{x^5}{120} + \dfrac{x^6}{720} + \cdots\right)$

$\qquad\qquad + \left(1 - x + \dfrac{x^2}{2} - \dfrac{x^3}{6} + \dfrac{x^4}{24} - \dfrac{x^5}{120} + \dfrac{x^6}{720} - \cdots\right)$

$\qquad\qquad = 2 + x^2 + \dfrac{x^4}{12} + \dfrac{x^6}{360} + \cdots$

$\dfrac{1}{e^x + e^{-x}} = \dfrac{1}{2 + x^2 + \dfrac{x^4}{12} + \dfrac{x^6}{360} + \cdots} = \dfrac{1}{2} - \dfrac{x^2}{4} + \dfrac{5x^4}{48} - \dfrac{61x^6}{1440} + \cdots$

21. Separating variables we obtain

$$\dfrac{dy}{y} = -dx \implies \ln|y| = -x + c \implies y = c_1 e^{-x}.$$

Substituting $y = \sum_{n=0}^{\infty} c_n x^n$ into the differential equation leads to

$$y' + y = \underbrace{\sum_{n=1}^{\infty} n c_n x^{n-1}}_{k=n-1} + \underbrace{\sum_{n=0}^{\infty} c_n x^n}_{k=n} = \sum_{k=0}^{\infty}(k+1)c_{k+1}x^k + \sum_{k=0}^{\infty} c_k x^k = \sum_{k=0}^{\infty}[(k+1)c_{k+1} + c_k]x^k = 0.$$

Thus

$$(k+1)c_{k+1} + c_k = 0$$

and

$$c_{k+1} = -\frac{1}{k+1}c_k, \quad k = 0, 1, 2, \ldots.$$

Iterating we find

$$c_1 = -c_0$$

$$c_2 = -\frac{1}{2}c_1 = \frac{1}{2}c_0$$

$$c_3 = -\frac{1}{3}c_2 = -\frac{1}{6}c_0$$

$$c_4 = -\frac{1}{4}c_3 = \frac{1}{24}c_0$$

and so on. Therefore

$$y = c_0 - c_0 x + \frac{1}{2}c_0 x^2 - \frac{1}{6}c_0 x^3 + \frac{1}{24}c_0 x^4 - \cdots = c_0\left[1 - x + \frac{1}{2}x^2 - \frac{1}{6}x^3 + \frac{1}{24}x^4 - \cdots\right]$$

$$= c_0 \sum_{n=0}^{\infty} \frac{1}{n!}(-x)^n = c_0 e^{-x}.$$

24. Separating variables we obtain

$$\frac{dy}{y} = -x^3 dx \implies \ln|y| = -\frac{1}{4}x^4 + c \implies y = c_1 e^{-x^4/4}.$$

Substituting $y = \sum_{n=0}^{\infty} c_n x^n$ into the differential equation leads to

$$y' + x^3 2y = \underbrace{\sum_{n=1}^{\infty} n c_n x^{n-1}}_{k=n-4} + \underbrace{\sum_{n=0}^{\infty} c_n x^{n+3}}_{k=n} = \sum_{k=-3}^{\infty}(k+4)c_{k+4}x^{k+3} - \sum_{k=0}^{\infty} c_k x^{k+3}$$

$$= c_1 + 2c_2 x + 3c_3 x^2 + \sum_{k=0}^{\infty}[(k+4)c_{k+4} + c_k]x^{k+2} = 0.$$

Thus

$$c_1 = c_2 = c_3 = 0,$$

$$(k+4)c_{k+4} + c_k = 0$$

and

$$c_{k+4} = -\frac{1}{k+4}c_k, \quad k = 0, 1, 2, \ldots .$$

Iterating we find

$$c_4 = -\frac{1}{4}c_0$$

$$c_5 = c_6 = c_7 = 0$$

$$c_8 = -\frac{1}{8}c_4 = \frac{1}{2} \cdot \frac{1}{4^2}c_0$$

$$c_9 = c_{10} = c_{11} = 0$$

$$c_{12} = -\frac{1}{12}c_8 = -\frac{1}{2 \cdot 3} \cdot \frac{1}{4^3}c_0$$

and so on. Therefore

$$y = c_0 - \frac{1}{4}c_0 x^4 + \frac{1}{2} \cdot \frac{1}{4^2}c_0 x^8 - \frac{1}{2 \cdot 3} \cdot \frac{1}{4^3}c_0 x^{12} + \cdots$$

$$= c_0 \left[1 - \frac{x^4}{4} + \frac{1}{2}\left(\frac{x^4}{4}\right)^2 - \frac{1}{2 \cdot 3}\left(\frac{x^4}{4}\right)^3 + \cdots \right] = c_0 \sum_{n=0}^{\infty} \frac{1}{n!}\left(\frac{-x^4}{4}\right)^n = c_0 e^{-x^4/4}.$$

27. The auxiliary equation is $m^2 + 1 = 0$, so $y = c_1 \cos x + c_2 \sin x$. Substituting $y = \sum\limits_{n=0}^{\infty} c_n x^n$ into the differential equation leads to

$$y'' + y = \underbrace{\sum_{n=2}^{\infty} n(n-1)c_n x^{n-2}}_{k=n-2} + \underbrace{\sum_{n=0}^{\infty} c_n x^n}_{k=n} = \sum_{k=0}^{\infty}(k+2)(k+1)c_{k+2}x^k + \sum_{k=0}^{\infty} c_k x^k$$

$$= \sum_{k=0}^{\infty} [(k+2)(k+1)c_{k+2} + c_k]x^k = 0.$$

Thus

$$(k+2)(k+1)c_{k+2} + c_k = 0$$

and

$$c_{k+2} = -\frac{1}{(k+2)(k+1)}c_k, \quad k = 0, 1, 2, \ldots .$$

Iterating we find

$$c_2 = -\frac{1}{2}c_0$$

$$c_3 = -\frac{1}{3 \cdot 2}c_1$$

$$c_4 = -\frac{1}{4 \cdot 3}c_2 = \frac{1}{4 \cdot 3 \cdot 2}c_0$$

$$c_5 = -\frac{1}{5 \cdot 4}c_3 = \frac{1}{5 \cdot 4 \cdot 3 \cdot 2}c_1$$

$$c_6 = -\frac{1}{6 \cdot 5}c_4 = -\frac{1}{6!}c_0$$

$$c_7 = -\frac{1}{7 \cdot 6}c_5 = -\frac{1}{7!}c_1$$

and so on. Therefore

$$y = c_0 + c_1 x - \frac{1}{2}c_0 x^2 - \frac{1}{3!}c_1 x^3 + \frac{1}{4!}c_0 x^4 + \frac{1}{5!}c_1 x^5 - \cdots$$

$$= c_0 \left[1 - \frac{1}{2}x^2 + \frac{1}{4!}x^4 - \cdots \right] + c_1 \left[1 - \frac{1}{3!}x^3 + \frac{1}{5!}x^5 - \cdots \right]$$

$$= c_0 \sum_{n=0}^{\infty} \frac{(-1)^n x^{2n}}{(2n)!} + c_1 \sum_{n=0}^{\infty} \frac{(-1)^n x^{2n+1}}{(2n+1)!} = c_0 \cos x + c_1 \sin x.$$

30. The auxiliary equation is $2m^2 + m = m(2m+1) = 0$, so $y = c_1 + c_2 e^{-x/2}$. Substituting $y = \sum_{n=0}^{\infty} c_n x^n$ into the differential equation leads to

$$2y'' + y' = 2 \underbrace{\sum_{n=2}^{\infty} n(n-1)c_n x^{n-2}}_{k=n-2} + \underbrace{\sum_{n=1}^{\infty} n c_n x^{n-1}}_{k=n-1}$$

$$= 2 \sum_{k=0}^{\infty} (k+2)(k+1)c_{k+2} x^k + \sum_{k=0}^{\infty} (k+1)c_{k+1} x^k$$

$$= \sum_{k=0}^{\infty} [2(k+2)(k+1)c_{k+2} + (k+1)c_{k+1}]x^k = 0.$$

Thus

$$2(k+2)(k+1)c_{k+2} + (k+1)c_{k+1} = 0$$

and

$$c_{k+2} = -\frac{1}{2(k+2)}c_{k+1}, \quad k = 0, 1, 2, \ldots.$$

Iterating we find

$$c_2 = -\frac{1}{2}\frac{1}{2}c_1$$

$$c_3 = -\frac{1}{2}\frac{1}{3}c_2 = \frac{1}{2^2}\frac{1}{3\cdot 2}c_1$$

$$c_4 = -\frac{1}{2}\frac{1}{4}c_3 = \frac{1}{2^3}\frac{1}{4!}c_1$$

and so on. Therefore

$$y = c_0 + c_1 x - \frac{1}{2}\frac{1}{2}c_1 x^2 + \frac{1}{2^2 3!}c_1 x^3 - \frac{1}{2^3 4!}c_1 x^4 + \cdots$$

$$\boxed{\begin{array}{l} c_0 = C_0 - 2c_1 \\ c_1 = -\frac{1}{2}C_1 \end{array}}$$

$$= C_0 + \left[C_1 - \frac{1}{2}C_1 x + \frac{1}{2}\frac{1}{2}\frac{1}{2}C_1 x^2 - \frac{1}{2^2 2\cdot 3!}\frac{1}{2}C_1 x^3 + \cdots\right]$$

$$= C_0 + C_1\left[1 - \frac{x}{2} + \frac{1}{2}\left(\frac{x}{2}\right)^2 - \frac{1}{3!}\left(\frac{x}{3}\right)^3 + \cdots\right]$$

$$= C_0 + C_1 \sum_{n=0}^{\infty} \frac{(-1)^n}{n!}\left(\frac{x}{2}\right)^n = C_0 + C_1 \sum_{n=0}^{\infty} \frac{1}{n!}\left(-\frac{x}{n}\right)^n = C_0 + C_1 e^{-x/2}.$$

Exercises 6.3

3. Substituting $y = \sum_{n=0}^{\infty} c_n x^n$ into the differential equation we have

$$y'' - 2xy' + y = \underbrace{\sum_{n=2}^{\infty} n(n-1)c_n x^{n-2}}_{k=n-2} - 2\underbrace{\sum_{n=1}^{\infty} nc_n x^n}_{k=n} + \underbrace{\sum_{n=0}^{\infty} c_n x^n}_{k=n}$$

$$= \sum_{k=0}^{\infty}(k+2)(k+1)c_{k+2}x^k - 2\sum_{k=1}^{\infty} kc_k x^k + \sum_{k=0}^{\infty} c_k x^k$$

$$= 2c_2 + c_0 + \sum_{k=1}^{\infty}[(k+2)(k+1)c_{k+2} - (2k-1)c_k]x^k = 0.$$

Thus

$$2c_2 + c_0 = 0$$

$$(k+2)(k+1)c_{k+2} - (2k-1)c_k = 0$$

65

and

$$c_2 = -\frac{1}{2}c_0$$

$$c_{k+2} = \frac{2k-1}{(k+2)(k+1)}c_k, \quad k = 1, 2, 3, \ldots.$$

Choosing $c_0 = 1$ and $c_1 = 0$ we find

$$c_2 = -\frac{1}{2}$$

$$c_3 = c_5 = c_7 = \cdots = 0$$

$$c_4 = -\frac{1}{8}$$

$$c_6 = -\frac{7}{336}$$

and so on. For $c_0 = 0$ and $c_1 = 1$ we obtain

$$c_2 = c_4 = c_6 = \cdots = 0$$

$$c_3 = \frac{1}{6}$$

$$c_5 = \frac{1}{24}$$

$$c_7 = \frac{1}{112}$$

and so on. Thus, two solutions are

$$y_1 = 1 - \frac{1}{2}x^2 - \frac{1}{8}x^4 - \frac{7}{336}x^6 - \cdots \quad \text{and} \quad y_2 = x + \frac{1}{6}x^3 + \frac{1}{24}x^5 + \frac{1}{112}x^7 + \cdots.$$

6. Substituting $y = \sum_{n=0}^{\infty} c_n x^n$ into the differential equation we have

$$y'' + 2xy' + 2y = \underbrace{\sum_{n=2}^{\infty} n(n-1)c_n x^{n-2}}_{k=n-2} + 2\underbrace{\sum_{n=1}^{\infty} nc_n x^n}_{k=n} + 2\underbrace{\sum_{n=0}^{\infty} c_n x^n}_{k=n}$$

$$= \sum_{k=0}^{\infty} (k+2)(k+1)c_{k+2}x^k + 2\sum_{k=1}^{\infty} kc_k x^k + 2\sum_{k=0}^{\infty} c_k x^k$$

$$= 2c_2 + 2c_0 + \sum_{k=1}^{\infty} [(k+2)(k+1)c_{k+2} + 2(k+1)c_k]x^k = 0.$$

Thus

$$2c_2 + 2c_0 = 0$$

$$(k+2)(k+1)c_{k+2} + 2(k+1)c_k = 0$$

and

$$c_2 = -c_0$$

$$c_{k+2} = -\frac{2}{k+2} c_k, \quad k = 1, 2, 3, \dots .$$

Choosing $c_0 = 1$ and $c_1 = 0$ we find

$$c_2 = -1$$

$$c_3 = c_5 = c_7 = \cdots = 0$$

$$c_4 = \frac{1}{2}$$

$$c_6 = -\frac{1}{6}$$

and so on. For $c_0 = 0$ and $c_1 = 1$ we obtain

$$c_2 = c_4 = c_6 = \cdots = 0$$

$$c_3 = -\frac{2}{3}$$

$$c_5 = \frac{4}{15}$$

$$c_7 = -\frac{8}{105}$$

and so on. Thus, two solutions are

$$y_1 = 1 - x^2 + \frac{1}{2}x^4 - \frac{1}{6}x^6 + \cdots \quad \text{and} \quad y_2 = x - \frac{2}{3}x^3 + \frac{4}{15}x^5 - \frac{8}{105}x^7 + \cdots .$$

9. Substituting $y = \sum_{n=0}^{\infty} c_n x^n$ into the differential equation we have

$$\left(x^2 - 1\right)y'' + 4xy' + 2y = \underbrace{\sum_{n=2}^{\infty} n(n-1)c_n x^n}_{k=n} - \underbrace{\sum_{n=2}^{\infty} n(n-1)c_n x^{n-2}}_{k=n-2} + 4\underbrace{\sum_{n=1}^{\infty} nc_n x^n}_{k=n} + 2\underbrace{\sum_{n=0}^{\infty} c_n x^n}_{k=n}$$

$$= \sum_{k=2}^{\infty} k(k-1)c_k x^k - \sum_{k=0}^{\infty} (k+2)(k+1)c_{k+2} x^k + 4\sum_{k=1}^{\infty} kc_k x^k + 2\sum_{k=0}^{\infty} c_k x^k$$

$$= -2c_2 + 2c_0 + (-6c_3 + 6c_1)x + \sum_{k=2}^{\infty} \left[\left(k^2 - k + 4k + 2\right)c_k - (k+2)(k+1)c_{k+2}\right]x^k = 0.$$

Thus

$$-2c_2 + 2c_0 = 0$$

67

$$-6c_3 + 6c_1 = 0$$

$$\left(k^2 + 3k + 2\right)c_k - (k+2)(k+1)c_{k+2} = 0$$

and

$$c_2 = c_0$$

$$c_3 = c_1$$

$$c_{k+2} = c_k, \quad k = 2, 3, 4, \ldots.$$

Choosing $c_0 = 1$ and $c_1 = 0$ we find

$$c_2 = 1$$

$$c_3 = c_5 = c_7 = \cdots = 0$$

$$c_4 = c_6 = c_8 = \cdots = 1.$$

For $c_0 = 0$ and $c_1 = 1$ we obtain

$$c_2 = c_4 = c_6 = \cdots = 0$$

$$c_3 = c_5 = c_7 = \cdots = 1.$$

Thus, two solutions are

$$y_1 = 1 + x^2 + x^4 + \cdots \quad \text{and} \quad y_2 = x + x^3 + x^5 + \cdots.$$

12. Substituting $y = \sum_{n=0}^{\infty} c_n x^n$ into the differential equation we have

$$\left(x^2 - 1\right)y'' + xy' - y = \underbrace{\sum_{n=2}^{\infty} n(n-1)c_n x^n}_{k=n} - \underbrace{\sum_{n=2}^{\infty} n(n-1)c_n x^{n-2}}_{k=n-2} + \underbrace{\sum_{n=1}^{\infty} nc_n x^n}_{k=n} - \underbrace{\sum_{n=0}^{\infty} c_n x^n}_{k=n}$$

$$= \sum_{k=2}^{\infty} k(k-1)c_k x^k - \sum_{k=0}^{\infty} (k+2)(k+1)c_{k+2} x^k + \sum_{k=1}^{\infty} kc_k x^k - \sum_{k=0}^{\infty} c_k x^k$$

$$= (-c_2 - c_0) - 6c_3 x + \sum_{k=2}^{\infty} \left[-(k+2)(k+1)c_{k+2} + \left(k^2 - 1\right)c_k\right]x^k = 0.$$

Thus

$$-2c_2 - c_0 = 0$$

$$-6c_3 = 0$$

$$-(k+2)(k+1)c_{k+2} + (k-1)(k+1)c_k = 0$$

and

$$c_2 = -\frac{1}{2}c_0$$

$$c_3 = 0$$

$$c_{k+2} = \frac{k-1}{k+2}c_k, \quad k = 2, 3, 4, \ldots.$$

Choosing $c_0 = 1$ and $c_1 = 0$ we find

$$c_2 = -\frac{1}{2}$$

$$c_3 = c_5 = c_7 = \cdots = 0$$

$$c_4 = -\frac{1}{8}$$

and so on. For $c_0 = 0$ and $c_1 = 1$ we obtain

$$c_2 = c_4 = c_6 = \cdots = 0$$

$$c_3 = c_5 = c_7 = \cdots = 0.$$

Thus, two solutions are

$$y_1 = 1 - \frac{1}{2}x^2 - \frac{1}{8}x^4 - \cdots \quad \text{and} \quad y_2 = x.$$

15. Substituting $y = \sum_{n=0}^{\infty} c_n x^n$ into the differential equation we have

$$(x-1)y'' - xy' + y = \underbrace{\sum_{n=2}^{\infty} n(n-1)c_n x^{n-1}}_{k=n-1} - \underbrace{\sum_{n=2}^{\infty} n(n-1)c_n x^{n-2}}_{k=n-2} - \underbrace{\sum_{n=1}^{\infty} nc_n x^n}_{k=n} + \underbrace{\sum_{n=0}^{\infty} c_n x^n}_{k=n}$$

$$= \sum_{k=1}^{\infty}(k+1)kc_{k+1}x^k - \sum_{k=0}^{\infty}(k+2)(k+1)c_{k+2}x^k - \sum_{k=1}^{\infty}kc_k x^k + \sum_{k=0}^{\infty}c_k x^k$$

$$= -2c_2 + c_0 + \sum_{k=1}^{\infty}[-(k+2)(k+1)c_{k+2} + (k+1)kc_{k+1} - (k-1)c_k]x^k = 0.$$

Thus

$$-2c_2 + c_0 = 0$$

$$-(k+2)(k+1)c_{k+2} + (k-1)kc_{k+1} - (k-1)c_k = 0$$

and

$$c_2 = \frac{1}{2}c_0$$

$$c_{k+2} = \frac{kc_{k+1}}{k+2} - \frac{(k-1)c_k}{(k+2)(k+1)}, \quad k = 1, 2, 3, \ldots.$$

Choosing $c_0 = 1$ and $c_1 = 0$ we find

$$c_2 = \frac{1}{2}, \qquad c_3 = \frac{1}{6}, \qquad c_4 = 0$$

and so on. For $c_0 = 0$ and $c_1 = 1$ we obtain $c_2 = c_3 = c_4 = \cdots = 0$. Thus,

$$y = C_1 \left(1 + \frac{1}{2}x^2 + \frac{1}{6}x^3 + \cdots \right) + C_2 x$$

and

$$y' = C_1 \left(x + \frac{1}{2}x^2 + \cdots \right) + C_2.$$

The initial conditions imply $C_1 = -2$ and $C_2 = 6$, so

$$y = -2 \left(1 + \frac{1}{2}x^2 + \frac{1}{6}x^3 + \cdots \right) + 6x = 8x - 2e^x.$$

18. Substituting $y = \sum_{n=0}^{\infty} c_n x^n$ into the differential equation we have

$$(x^2 + 1)y'' + 2xy' = \underbrace{\sum_{n=2}^{\infty} n(n-1)c_n x^n}_{k=n} + \underbrace{\sum_{n=2}^{\infty} n(n-1)c_n x^{n-2}}_{k=n-2} + \underbrace{\sum_{n=1}^{\infty} 2nc_n x^n}_{k=n}$$

$$= \sum_{k=2}^{\infty} k(k-1)c_k x^k + \sum_{k=0}^{\infty} (k+2)(k+1)c_{k+2} x^k + \sum_{k=1}^{\infty} 2kc_k x^k$$

$$= 2c_2 + (6c_3 + 2c_1)x + \sum_{k=2}^{\infty} [k(k+1)c_k + (k+2)(k+1)c_{k+2}]x^k = 0.$$

Thus

$$2c_2 = 0,$$

$$6c_3 + 2c_1 = 0,$$

$$k(k+1)c_k + (k+2)(k+1)c_{k+2} = 0$$

and

$$c_2 = 0$$

$$c_3 = -\frac{1}{3}c_1$$

$$c_{k+2} = -\frac{k}{k+2}c_k, \quad k = 2, 3, 4, \ldots.$$

Choosing $c_0 = 1$ and $c_1 = 0$ we find $c_3 = c_4 = c_5 = \cdots = 0$. For $c_0 = 0$ and $c_1 = 1$ we obtain

$$c_3 - \frac{1}{3}$$

$$c_4 = c_6 = c_8 = \cdots = 0$$

$$c_5 = -\frac{1}{5}$$

$$c_7 = \frac{1}{7}$$

and so on. Thus

$$y = c_0 + c_1 \left(x - \frac{1}{3}x^3 + \frac{1}{5}x^5 - \frac{1}{7}x^7 + \cdots \right)$$

and

$$y' = c_1 \left(1 - x^2 + x^4 - x^6 + \cdots \right).$$

The initial conditions imply $c_0 = 0$ and $c_1 = 1$, so

$$y = x - \frac{1}{3}x^3 + \frac{1}{5}x^5 - \frac{1}{7}x^7 + \cdots.$$

21. Substituting $y = \sum_{n=0}^{\infty} c_n x^n$ into the differential equation we have

$$y'' + e^{-x}y = \sum_{n=2}^{\infty} n(n-1)c_n x^{n-2}$$

$$+ \left(1 - x + \frac{1}{2}x^2 - \frac{1}{6}x^3 + \frac{1}{24}x^4 - \cdots \right) \left(c_0 + c_1 x + c_2 x^2 + c_3 x^3 + \cdots \right)$$

$$= \left[2c_2 + 6c_3 x + 12c_4 x^2 + 20c_5 x^3 + \cdots \right] + \left[c_0 + (c_1 - c_0)x + \left(c_2 - c_1 + \frac{1}{2}c_0 \right)x^2 + \cdots \right]$$

$$= (2c_2 + c_0) + (6c_3 + c_1 - c_0)x + (12c_4 + c_2 - c_1 + \frac{1}{2}c_0)x^2 + \cdots = 0.$$

Then

$$2c_2 + c_0 = 0$$

$$6c_3 + c_1 - c_0 = 0$$

$$12c_4 + c_2 - c_1 + \frac{1}{2}c_0 = 0$$

and

$$c_2 = -\frac{1}{2}c_0$$

$$c_3 = -\frac{1}{6}c_1 + \frac{1}{6}c_0$$

$$c_4 = -\frac{1}{12}c_2 + \frac{1}{12}c_1 - \frac{1}{24}c_0.$$

Choosing $c_0 = 1$ and $c_1 = 0$ we find

$$c_2 = -\frac{1}{2}, \qquad c_3 = \frac{1}{6}, \qquad c_4 = 0$$

and so on. For $c_0 = 0$ and $c_1 = 1$ we obtain

$$c_2 = 0, \qquad c_3 = -\frac{1}{6}, \qquad c_4 = \frac{1}{12}.$$

Thus, two solutions are

$$y_1 = 1 - \frac{1}{2}x^2 + \frac{1}{6}x^3 + \cdots \quad \text{and} \quad y_2 = x - \frac{1}{6}x^3 + \frac{1}{12}x^4 + \cdots.$$

24. Substituting $y = \sum_{n=0}^{\infty} c_n x^n$ into the differential equation leads to

$$y'' - 4xy' - 4y = \underbrace{\sum_{n=2}^{\infty} n(n-1)c_n x^{n-2}}_{k=n-2} - \underbrace{\sum_{n=1}^{\infty} 4nc_n x^n}_{k=n} - \underbrace{\sum_{n=0}^{\infty} 4c_n x^n}_{k=n}$$

$$= \sum_{k=0}^{\infty} (k+2)(k+1)c_{k+2}x^k - \sum_{k=1}^{\infty} 4kc_k x^k - \sum_{k=0}^{\infty} 4c_k x^k$$

$$= 2c_2 - 4c_0 + \sum_{k=1}^{\infty} [(k+2)(k+1)c_{k+2} - 4(k+1)c_k]x^k$$

$$= e^x = 1 + \sum_{k=1}^{\infty} \frac{1}{k!}x^k.$$

Thus

$$2c_2 - 4c_0 = 1$$

$$(k+2)(k+1)c_{k+2} - 4(k+1)c_k = \frac{1}{k!}$$

and

$$c_2 = \frac{1}{2} + 2c_0$$

$$c_{k+2} = \frac{1}{(k+2)!} + \frac{4}{k+2}c_k, \qquad k = 1, 2, 3, \ldots.$$

Let c_0 and c_1 be arbitrary and iterate to find

$$c_2 = \frac{1}{2} + 2c_0$$

$$c_3 = \frac{1}{3!} + \frac{4}{3}c_1 = \frac{1}{3!} + \frac{4}{3}c_1$$

$$c_4 = \frac{1}{4!} + \frac{4}{4}c_2 = \frac{1}{4!} + \frac{1}{2} + 2c_0 = \frac{13}{4!} + 2c_0$$

$$c_5 = \frac{1}{5!} + \frac{4}{5}c_3 = \frac{1}{5!} + \frac{4}{5 \cdot 3!} + \frac{16}{15}c_1 = \frac{17}{5!} + \frac{16}{15}c_1$$

$$c_6 = \frac{1}{6!} + \frac{4}{6}c_4 = \frac{1}{6!} + \frac{4 \cdot 13}{6 \cdot 4!} + \frac{8}{6}c_0 = \frac{261}{6!} + \frac{4}{3}c_0$$

$$c_7 = \frac{1}{7!} + \frac{4}{7}c_5 = \frac{1}{7!} + \frac{4 \cdot 17}{7 \cdot 5!} + \frac{64}{105}c_1 = \frac{409}{7!} + \frac{64}{105}c_1$$

and so on. The solution is

$$y = c_0 + c_1 x + \left(\frac{1}{2} + 2c_0\right)x^2 + \left(\frac{1}{3!} + \frac{4}{3}c_1\right)x^3 - \left(\frac{13}{4!} + 2c_0\right)x^4 + \left(\frac{17}{5!} + \frac{16}{15}c_1\right)x^5$$

$$+ \left(\frac{261}{6!} + \frac{4}{3}c_0\right)x^6 + \left(\frac{409}{7!} + \frac{64}{105}c_1\right)x^7 + \cdots$$

$$= c_0\left[1 + 2x^2 + 2x^4 + \frac{4}{3}x^6 + \cdots\right] + c_1\left[x + \frac{4}{3}x^3 + \frac{16}{15}x^5 + \frac{64}{105}x^7 + \cdots\right]$$

$$+ \frac{1}{2}x^2 + \frac{1}{3!}x^3 + \frac{13}{4!}x^4 + \frac{17}{5!}x^5 + \frac{261}{6!}x^6 + \frac{409}{7!}x^7 + \cdots.$$

Exercises 6.4

3. Irregular singular point: $x = 3$. Regular singular point: $x = -3$.

6. Irregular singular point: $x = 5$. Regular singular point: $x = 0$.

9. Irregular singular point: $x = 0$. Regular singular points: $x = 2, \pm 5$.

12. Substituting $y = \sum_{n=0}^{\infty} c_n x^{n+r}$ into the differential equation and collecting terms, we obtain

$$2xy'' + 5y' + xy = \left(2r^2 + 3r\right)c_0 x^{r-1} + \left(2r^2 + 7r + 5\right)c_1 x^r$$

$$+ \sum_{k=2}^{\infty} [2(k+r)(k+r-1)c_k + 5(k+r)c_k + c_{k-2}]x^{k+r-1}$$

$$= 0,$$

which implies

$$2r^2 + 3r = r(2r + 3) = 0,$$

$$\left(2r^2 + 7r + 5\right) c_1 = 0,$$

and

$$(k+r)(2k+2r+3)c_k + c_{k-2} = 0.$$

The indicial roots are $r = -3/2$ and $r = 0$, so $c_1 = 0$. For $r = -3/2$ the recurrence relation is

$$c_k = -\frac{c_{k-2}}{(2k-3)k}, \quad k = 2, 3, 4, \dots,$$

and

$$c_2 = -\frac{1}{2}c_0, \qquad c_3 = 0, \qquad c_4 = \frac{1}{40}c_0.$$

For $r = 0$ the recurrence relation is

$$c_k = -\frac{c_{k-2}}{k(2k+3)}, \quad k = 2, 3, 4, \dots,$$

and

$$c_2 = -\frac{1}{14}c_0, \qquad c_3 = 0, \qquad c_4 = \frac{1}{616}c_0.$$

The general solution on $(0, \infty)$ is

$$y = C_1 x^{-3/2} \left(1 - \frac{1}{2}x^2 + \frac{1}{40}x^4 + \cdots\right) + C_2 \left(1 - \frac{1}{14}x^2 + \frac{1}{616}x^4 + \cdots\right).$$

15. Substituting $y = \sum_{n=0}^{\infty} c_n x^{n+r}$ into the differential equation and collecting terms, we obtain

$$3xy'' + (2-x)y' - y = \left(3r^2 - r\right) c_0 x^{r-1}$$

$$+ \sum_{k=1}^{\infty} [3(k+r-1)(k+r)c_k + 2(k+r)c_k - (k+r)c_{k-1}]x^{k+r-1}$$

$$= 0,$$

which implies

$$3r^2 - r = r(3r - 1) = 0$$

and

$$(k+r)(3k+3r-1)c_k - (k+r)c_{k-1} = 0.$$

The indicial roots are $r = 0$ and $r = 1/3$. For $r = 0$ the recurrence relation is

$$c_k = \frac{c_{k-1}}{(3k-1)}, \quad k = 1, 2, 3, \dots,$$

and

$$c_1 = \frac{1}{2}c_0, \qquad c_2 = \frac{1}{10}c_0, \qquad c_3 = \frac{1}{80}c_0.$$

For $r = 1/3$ the recurrence relation is

$$c_k = \frac{c_{k-1}}{3k}, \quad k = 1, 2, 3, \ldots,$$

and

$$c_1 = \frac{1}{3}c_0, \qquad c_2 = \frac{1}{18}c_0, \qquad c_3 = \frac{1}{162}c_0.$$

The general solution on $(0, \infty)$ is

$$y = C_1 \left(1 + \frac{1}{2}x + \frac{1}{10}x^2 + \frac{1}{80}x^3 + \cdots\right) + C_2 x^{1/3}\left(1 + \frac{1}{3}x + \frac{1}{18}x^2 + \frac{1}{162}x^3 + \cdots\right).$$

18. Substituting $y = \sum_{n=0}^{\infty} c_n x^{n+r}$ into the differential equation and collecting terms, we obtain

$$2xy'' + xy' + \left(x^2 - \frac{4}{9}\right)y = \left(r^2 - \frac{4}{9}\right)c_0 x^r + \left(r^2 + 2r + \frac{5}{9}\right)c_1 x^{r+1}$$

$$+ \sum_{k=2}^{\infty}\left[(k+r)(k+r-1)c_k + (k+r)c_k - \frac{4}{9}c_k + c_{k-2}\right]x^{k+r}$$

$$= 0,$$

which implies

$$r^2 - \frac{4}{9} = \left(r + \frac{2}{3}\right)\left(r - \frac{2}{3}\right) = 0,$$

$$\left(r^2 + 2r + \frac{5}{9}\right)c_1 = 0,$$

and

$$\left[(k+r)^2 - \frac{4}{9}\right]c_k + c_{k-2} = 0.$$

The indicial roots are $r = -2/3$ and $r = 2/3$, so $c_1 = 0$. For $r = -2/3$ the recurrence relation is

$$c_k = -\frac{9c_{k-2}}{3k(3k-4)}, \quad k = 2, 3, 4, \ldots,$$

and

$$c_2 = -\frac{3}{4}c_0, \qquad c_3 = 0, \qquad c_4 = \frac{9}{128}c_0.$$

For $r = 2/3$ the recurrence relation is

$$c_k = -\frac{9c_{k-2}}{3k(3k+4)}, \quad k = 2, 3, 4, \ldots,$$

and

$$c_2 = -\frac{3}{20}c_0, \qquad c_3 = 0, \qquad c_4 = \frac{9}{1,280}c_0.$$

The general solution on $(0, \infty)$ is

$$y = C_1 x^{-2/3}\left(1 - \frac{3}{4}x^2 + \frac{9}{128}x^4 + \cdots\right) + C_2 x^{2/3}\left(1 - \frac{3}{20}x^2 + \frac{9}{1,280}x^4 + \cdots\right).$$

21. Substituting $y = \sum_{n=0}^{\infty} c_n x^{n+r}$ into the differential equation and collecting terms, we obtain

$$2x^2 y'' - x(x-1)y' - y = \left(2r^2 - r - 1\right) c_0 x^r$$

$$+ \sum_{k=1}^{\infty} [2(k+r)(k+r-1)c_k + (k+r)c_k - c_k - (k+r-1)c_{k-1}]x^{k+r}$$

$$= 0,$$

which implies
$$2r^2 - r - 1 = (2r+1)(r-1) = 0$$

and
$$[(k+r)(2k+2r-1) - 1]c_k - (k+r-1)2c_{k-1} = 0.$$

The indicial roots are $r = -1/2$ and $r = 1$. For $r = -1/2$ the recurrence relation is

$$c_k = \frac{c_{k-1}}{2k}, \quad k = 1, 2, 3, \ldots,$$

and
$$c_1 = \frac{1}{2}c_0, \quad c_2 = \frac{1}{8}c_0, \quad c_3 = \frac{1}{48}c_0.$$

For $r = 1$ the recurrence relation is

$$c_k = \frac{c_{k-1}}{2k+3}, \quad k = 1, 2, 3, \ldots,$$

and
$$c_1 = \frac{1}{5}c_0, \quad c_2 = \frac{1}{35}c_0, \quad c_3 = \frac{1}{315}c_0.$$

The general solution on $(0, \infty)$ is

$$y = C_1 x^{-1/2}\left(1 + \frac{1}{2}x + \frac{1}{8}x^2 + \frac{1}{48}x^3 + \cdots\right) + C_2 x\left(1 + \frac{1}{5}x + \frac{1}{35}x^2 + \frac{1}{315}x^3 + \cdots\right).$$

24. Substituting $y = \sum_{n=0}^{\infty} c_n x^{n+r}$ into the differential equation and collecting terms, we obtain

$$x^2 y'' + xy' + \left(x^2 - \frac{1}{4}\right)y = \left(r^2 - \frac{1}{4}\right)c_0 x^r + \left(r^2 + 2r + \frac{3}{4}\right)c_1 x^{r+1}$$

$$+ \sum_{k=2}^{\infty} [(k+r)(k+r-1)c_k + (k+r)c_k - \frac{1}{4}c_k + c_{k-2}]x^{k+r}$$

$$= 0,$$

which implies

$$r^2 - \frac{1}{4} = \left(r - \frac{1}{2}\right)\left(r + \frac{1}{2}\right) = 0,$$

$$\left(r^2 + 2r + \frac{3}{4}\right)c_1 = 0,$$

and

$$\left[(k+r)^2 - \frac{1}{4}\right] c_k + c_{k-2} = 0.$$

The indicial roots are $r_1 = 1/2$ and $r_2 = -1/2$, so $c_1 = 0$. For $r_1 = 1/2$ the recurrence relation is

$$c_k = -\frac{c_{k-2}}{k(k+1)}, \quad k = 2, 3, 4, \ldots,$$

and

$$c_2 = -\frac{1}{3!} c_0$$

$$c_3 = c_5 = c_7 = \cdots = 0$$

$$c_4 = \frac{1}{5!} c_0$$

$$c_{2n} = \frac{(-1)^n}{(2n+1)!} c_0.$$

For $r_2 = -1/2$ the recurrence relation is

$$c_k = -\frac{c_{k-2}}{k(k-1)}, \quad k = 2, 3, 4, \ldots,$$

and

$$c_2 = -\frac{1}{2!} c_0$$

$$c_3 = c_5 = c_7 = \cdots = 0$$

$$c_4 = \frac{1}{4!} c_0$$

$$c_{2n} = \frac{(-1)^n}{(2n)!} c_0.$$

The general solution on $(0, \infty)$ is

$$y = C_1 x^{1/2} \sum_{n=0}^{\infty} \frac{(-1)^n}{(2n+1)!} x^{2n} + C_2 x^{-1/2} \sum_{n=0}^{\infty} \frac{(-1)^n}{(2n)!} x^{2n}$$

$$= C_1 x^{-1/2} \sum_{n=0}^{\infty} \frac{(-1)^n}{(2n+1)!} x^{2n+1} + C_2 x^{-1/2} \sum_{n=0}^{\infty} \frac{(-1)^n}{(2n)!} x^{2n}$$

$$= x^{-1/2}[C_1 \sin x + C_2 \cos x].$$

27. Substituting $y = \sum_{n=0}^{\infty} c_n x^{n+r}$ into the differential equation and collecting terms, we obtain

$$xy'' + (1-x)y' - y = r^2 c_0 x^{r-1} + \sum_{k=0}^{\infty} [(k+r)(k+r-1)c_k + (k+r)c_k - (k+r)c_{k-1}] x^{k+r-1} = 0,$$

which implies $r^2 = 0$ and

$$(k+r)^2 c_k - (k+r)c_{k-1} = 0.$$

The indicial roots are $r_1 = r_2 = 0$ and the recurrence relation is

$$c_k = \frac{c_{k-1}}{k}, \quad k = 1, 2, 3, \ldots .$$

One solution is

$$y_1 = c_0 \left(1 + x + \frac{1}{2}x^2 + \frac{1}{3!}x^3 + \cdots \right) = c_0 e^x.$$

A second solution is

$$y_2 = y_1 \int \frac{e^{-\int (1/x - 1)dx}}{e^{2x}} \, dx = e^x \int \frac{e^x/x}{e^{2x}} \, dx = e^x \int \frac{1}{x} e^{-x} dx$$

$$= e^x \int \frac{1}{x} \left(1 - x + \frac{1}{2}x^2 - \frac{1}{3!}x^3 + \cdots \right) dx = e^x \int \left(\frac{1}{x} - 1 + \frac{1}{2}x - \frac{1}{3!}x^2 + \cdots \right) dx$$

$$= e^x \left[\ln x - x + \frac{1}{2 \cdot 2}x^2 - \frac{1}{3 \cdot 3!}x^3 + \cdots \right] = e^x \ln x - e^x \sum_{n=1}^{\infty} \frac{(-1)^{n+1}}{n \cdot n!} x^n.$$

The general solution on $(0, \infty)$ is

$$y = C_1 e^x + C_2 e^x \left(\ln x - \sum_{n=1}^{\infty} \frac{(-1)^{n+1}}{n \cdot n!} x^n \right).$$

30. Substituting $y = \sum_{n=0}^{\infty} c_n x^{n+r}$ into the differential equation and collecting terms, we obtain

$$xy'' - xy' + y = \left(r^2 - r \right) c_0 x^{r-1} + \sum_{k=0}^{\infty} [(k + r + 1)(k + r)c_{k+1} - (k + r)c_k + c_k] x^{k+r} = 0$$

which implies

$$r^2 - r = r(r - 1) = 0$$

and

$$(k + r + 1)(k + r)c_{k+1} - (k + r - 1)c_k = 0.$$

The indicial roots are $r_1 = 1$ and $r_2 = 0$. For $r_1 = 1$ the recurrence relation is

$$c_{k+1} = \frac{kc_k}{(k + 2)(k + 1)}, \quad k = 0, 1, 2, \ldots,$$

and one solution is $y_1 = c_0 x$. A second solution is

$$y_2 = x \int \frac{e^{-\int -dx}}{x^2} \, dx = x \int \frac{e^x}{x^2} \, dx = x \int \frac{1}{x^2} \left(1 + x + \frac{1}{2}x^2 + \frac{1}{3!}x^3 + \cdots \right) dx$$

$$= x \int \left(\frac{1}{x^2} + \frac{1}{x} + \frac{1}{2} + \frac{1}{3!}x + \frac{1}{4!}x^2 + \cdots \right) dx = x \left[-\frac{1}{x} + \ln x + \frac{1}{2}x + \frac{1}{12}x^2 + \frac{1}{72}x^3 + \cdots \right]$$

$$= x \ln x - 1 + \frac{1}{2}x^2 + \frac{1}{12}x^3 + \frac{1}{72}x^4 + \cdots .$$

The general solution on $(0, \infty)$ is

$$y = C_1 x + C_2 y_2(x).$$

33. Substituting $y = \sum_{n=0}^{\infty} c_n x^{n+r}$ into the differential equation and collecting terms, we obtain

$$xy'' + (x-1)y' - 2y = r^2 c_0 x^{r-1} + \sum_{k=1}^{\infty} [(k+r)(k+r-1)c_k$$

$$- (k+r)c_k + (k+r-3)c_{k-1}]x^{k+r-1}$$

$$= 0$$

which implies $r^2 = 0$ and

$$(k+r)(k+r-2)c_k + (k+r-3)c_{k-1} = 0.$$

The indicial roots are $r_1 = r_2 = 0$ and the recurrence relation is

$$k(k-2)c_k + (k-3)c_{k-1} = 0, \quad k = 1, 2, 3, \dots .$$

Then

$$-c_1 - 2c_0 = 0 \implies c_1 = -2c_0$$

$$0c_2 - c_1 = 0 \implies c_1 = 0 \text{ and } c_2 \text{ is arbitrary}$$

$$3c_3 + 0c_2 = 0 \implies c_3 = 0$$

and

$$c_k = -\frac{(k-3)c_{k-1}}{k(k-2)}, \quad k = 4, 5, 6, \dots .$$

Since $c_1 = 0$ and $c_1 = -2c_0$, we have $c_0 = 0$. Taking $c_2 = 0$ we obtain $c_3 = c_4 = c_5 = \cdots = 0$. Thus, $y_1 = c_2 x^2$. A second solution is

$$y_2 = x^2 \int \frac{e^{-\int (1-1/x)\,dx}}{x^4}\,dx = x^2 \int \frac{xe^{-x}}{x^4}\,dx = x^2 \int \frac{1}{x^3}\left(1 - x + \frac{1}{2}x^2 - \frac{1}{3!}x^3 + \frac{1}{4!}x^4 - \cdots\right)dx$$

$$= x^2 \int \left(\frac{1}{x^3} - \frac{1}{x^2} + \frac{1}{2x} - \frac{1}{3!} + \frac{1}{4!}x - \cdots\right)dx = x^2\left[-\frac{1}{2x^2} + \frac{1}{x} + \frac{1}{2}\ln x - \frac{1}{6}x + \frac{1}{48}x^2 - \cdots\right]$$

$$= \frac{1}{2}x^2 \ln x - \frac{1}{2} + x - \frac{1}{6}x^3 + \frac{1}{48}x^4 - \cdots .$$

36. Substituting $y = \sum_{n=0}^{\infty} c_n x^{n+r}$ into the differential equation and collecting terms, we obtain

$$x^2 y'' - y' + y = rc_0 x^{r-1} + \sum_{k=0}^{\infty} ([\,(k+r)(k+r-1)+1]c_k - (k+r+1)c_{k+1})\, x^{k+r} = 0.$$

Thus $r = 0$ and the recurrence relation is

$$c_{k+1} = \frac{k(k-1)+1}{k+1}c_k, \quad k = 0, 1, 2, \dots .$$

Then $$c_1 = 0, \qquad c_2 = \frac{1}{2}c_0, \qquad c_3 = \frac{1}{2}c_0, \qquad c_4 = \frac{7}{8}c_0,$$

and so on. Therefore, one solution is

$$y(x) = c_0 \left[1 + x + \frac{1}{2}x^2 + \frac{1}{2}x^3 + \frac{7}{8}x^4 + \cdots \right].$$

39. Identifying $p_0 = 5/3$ and $q_0 = -1/3$, the indicial equation is

$$r(r-1) + \frac{5}{3}r - \frac{1}{3} = r^2 + \frac{2}{3}r - \frac{1}{3} = (r+1)\left(r - \frac{1}{3}\right) = 0.$$

The indicial roots are -1 and $1/3$.

Exercises 6.5

3. Since $\nu^2 = 25/4$ the general solution is $y = c_1 J_{5/2}(x) + c_2 J_{-5/2}(x)$.

6. Since $\nu^2 = 4$ the general solution is $y = c_1 J_2(x) + c_2 Y_2(x)$.

9. If $y = x^{-1/2}v(x)$ then

$$y' = x^{-1/2}v'(x) - \frac{1}{2}x^{-3/2}v(x),$$

$$y'' = x^{-1/2}v''(x) - x^{-3/2}v'(x) + \frac{3}{4}x^{-5/2}v(x),$$

and

$$x^2 y'' + 2xy' + \lambda^2 x^2 y = x^{3/2}v'' + x^{1/2}v' + \left(\lambda^2 x^{3/2} - \frac{1}{4}x^{-1/2}\right)v.$$

Multiplying by $x^{1/2}$ we obtain

$$x^2 v'' + xv' + \left(\lambda^2 x^2 - \frac{1}{4}\right)v = 0,$$

whose solution is $v = c_1 J_{1/2}(\lambda x) + c_2 J_{-1/2}(\lambda x)$. Then $y = c_1 x^{-1/2} J_{1/2}(\lambda x) + c_2 x^{-1/2} J_{-1/2}(\lambda x)$.

12. From $y = \sqrt{x} J_\nu(\lambda x)$ we find

$$y' = \lambda \sqrt{x} J_\nu'(\lambda x) + \frac{1}{2}x^{-1/2} J_\nu(\lambda x)$$

and

$$y'' = \lambda^2 \sqrt{x} J_\nu''(\lambda x) + \lambda x^{-1/2} J_\nu'(\lambda x) - \frac{1}{4}x^{-3/2} J_\nu(\lambda x).$$

Substituting into the differential equation, we have

$$x^2 y'' + \left(\lambda^2 x^2 - \nu^2 + \frac{1}{4}\right)y = \sqrt{x}\left[\lambda^2 x^2 J_\nu''(\lambda x) + \lambda x J_\nu'(\lambda x) + \left(\lambda^2 x^2 - \nu^2\right)J_\nu(\lambda x)\right]$$

$$= \sqrt{x} \cdot 0 \qquad \text{(since } J_n \text{ is a solution of Bessel's equation)}$$

$$= 0.$$

Therefore, $\sqrt{x}\,J_\nu(\lambda x)$ is a solution of the original equation.

15. From Problem 10 with $n = -1$ we find $y = x^{-1}J_{-1}(x)$. From Problem 11 with $n = 1$ we find $y = x^{-1}J_1(x) = -x^{-1}J_{-1}(x)$.

18. From Problem 10 with $n = 3$ we find $y = x^3J_3(x)$. From Problem 11 with $n = -3$ we find $y = x^3J_{-3}(x) = -x^3J_3(x)$.

21. The recurrence relation follows from

$$xJ_{\nu+1}(x) + xJ_{\nu-1}(x) = \sum_{n=0}^{\infty} \frac{(-1)^{n-1}2n}{n!\Gamma(1+\nu+n)}\left(\frac{x}{2}\right)^{2n+\nu} + \sum_{n=0}^{\infty}\frac{(-1)^n 2(\nu+n)}{n!\Gamma(1+\nu+n)}\left(\frac{x}{2}\right)^{2n+\nu}$$

$$= \sum_{n=0}^{\infty}\frac{(-1)^n 2\nu}{n!\Gamma(1+\nu+n)}\left(\frac{x}{2}\right)^{2n+\nu} = 2\nu J_\nu(x).$$

24. By Problem 19 we obtain $J_0'(x) = J_{-1}(x)$ and by Problem 22

$$2J_0'(x) = J_{-1}(x) - J_1(x) = J_0'(x) - J_1(x)$$

so that $J_0'(x) = -J_1(x)$.

27. Since

$$\Gamma\left(1 - \frac{1}{2} + n\right) = \frac{(2n-1)!}{(n-1)!2^{2n-1}}$$

we obtain

$$J_{-1/2}(x) = \sum_{n=0}^{\infty}\frac{(-1)^n 2^{1/2}x^{-1/2}}{2n(2n-1)!\sqrt{\pi}}x^{2n} = \sqrt{\frac{2}{\pi x}}\cos x.$$

30. By Problem 21 we obtain $3J_{3/2}(x) = xJ_{5/2}(x) + xJ_{1/2}(x)$ so that

$$J_{5/2}(x) = \sqrt{\frac{2}{\pi x}}\left(\frac{3\sin x}{x^2} - \frac{3\cos x}{x} - \sin x\right).$$

33. By Problem 21 we obtain $-5J_{-5/2}(x) = xJ_{-3/2}(x) + xJ_{-7/2}(x)$ so that

$$J_{-7/2}(x) = \sqrt{\frac{2}{\pi x}}\left(\frac{-15\cos x}{x^3} - \frac{15\sin x}{x^2} + \frac{6\cos x}{x} + \sin x\right).$$

36. If $y_1 = J_0(x)$ then using equation (35) on Page 299 in the text gives

$$y_2 = J_0(x) \int \frac{e^{-\int dx/x}}{(J_0(x))^2} \, dx$$

$$= J_0(x) \int \frac{dx}{x \left(1 - \dfrac{x^2}{4} + \dfrac{x^4}{64} - \dfrac{x^6}{2304} + \cdots \right)^2} \, dx$$

$$= J_0(x) \int \left(\frac{1}{x} + \frac{x}{2} + \frac{5x^3}{32} + \frac{23x^5}{576} + \cdots \right) dx$$

$$= J_0(x) \left(\ln x + \frac{x^2}{4} + \frac{5x^4}{128} + \frac{23x^6}{3456} + \cdots \right)$$

$$= J_0(x) \ln x + \left(1 - \frac{x^2}{4} + \frac{x^4}{64} - \frac{x^6}{2304} + \cdots \right)\left(\frac{x^2}{4} + \frac{5x^4}{128} + \frac{23x^6}{3456} + \cdots \right)$$

$$= J_0(x) \ln x + \frac{x^2}{4} - \frac{3x^4}{128} + \frac{11x^6}{13824} - \cdots .$$

39. (a) Using the formulas on Page 315 in the text we obtain

$$P_6(x) = \frac{1}{16}\left(231x^6 - 315x^4 + 105x^2 - 5\right)$$

and

$$P_7(x) = \frac{1}{16}\left(429x^7 - 693x^5 + 315x^3 - 35x\right).$$

(b) $P_6(x)$ satisfies $\left(1 - x^2\right)y'' - 2xy' + 42y = 0$ and $P_7(x)$ satisfies $\left(1 - x^2\right)y'' - 2xy' + 56y = 0$.

42. The polynomials are shown in (18) on Page 316 in the text.

45. The recurrence relation can be wrtten

$$P_{k+1}(x) = \frac{2k+1}{k+1}\, x P_k(x) - \frac{k}{k+1}\, P_{k-1}(x), \qquad k = 2,\ 3,\ 4,\ \dots .$$

$k = 1$: $\quad P_2(x) = \dfrac{3}{2}x^2 - \dfrac{1}{2}$

$k = 2$: $\quad P_3(x) = \dfrac{5}{3}x\left(\dfrac{3}{2}x^2 - \dfrac{1}{2}\right) - \dfrac{2}{3}x = \dfrac{5}{2}x^3 - \dfrac{3}{2}x$

$k = 3$: $\quad P_4(x) = \dfrac{7}{4}x\left(\dfrac{5}{2}x^3 - \dfrac{3}{2}x\right) - \dfrac{3}{4}\left(\dfrac{3}{2}x^2 - \dfrac{1}{2}\right) = \dfrac{35}{8}x^4 - \dfrac{30}{8}x^2 + \dfrac{3}{8}$

$k = 4$: $\quad P_5(x) = \dfrac{9}{5}x\left(\dfrac{35}{8}x^4 - \dfrac{30}{8}x^2 + \dfrac{3}{8}\right) - \dfrac{4}{5}\left(\dfrac{5}{2}x^3 - \dfrac{3}{2}x\right) = \dfrac{63}{8}x^5 - \dfrac{35}{4}x^3 + \dfrac{15}{8}x$

$k = 5$: $\quad P_6(x) = \dfrac{11}{6}x\left(\dfrac{63}{8}x^5 - \dfrac{35}{4}x^3 + \dfrac{15}{8}x\right) - \dfrac{5}{6}\left(\dfrac{35}{8}x^4 - \dfrac{30}{8}x^2 + \dfrac{3}{8}\right) = \dfrac{231}{16}x^6 - \dfrac{315}{16}x^4 + \dfrac{105}{16}x - \dfrac{5}{16}$

48. All integrals of the form $\int_{-1}^{1} P_n(x)P_m(x)\,dx$ are 0 for $n \neq m$.

Chapter 6 Review Exercises

3. The auxiliary equation is $m^2 - 5m + 6 = (m-2)(m-3) = 0$ and a particular solution is $y_p = x^4 - x^2 \ln x$ so that

$$y = c_1 x^2 + c_2 x^3 + x^4 - x^2 \ln x.$$

6. Since

$$P(x) = 0 \quad \text{and} \quad Q(x) = \frac{2}{(x^2-4)(x^2+4)}$$

the singular points are $x = 2$, $x = -2$, $x = 2i$, and $x = -2i$. All others are ordinary points.

9. Since

$$P(x) = -\frac{1}{x^2(x^2-9)} \quad \text{and} \quad Q(x) = \frac{1}{x(x^2-9)^2}$$

the regular singular points are $x = 3$ and $x = -3$. The irregular singular point is $x = 0$.

12. Since $P(x) = -2x/(x^2-4)$ and $Q(x) = 9/(x^2-4)$ an interval of convergence is $-2 < x < 2$.

15. Substituting $y = \sum_{n=0}^{\infty} c_n x^n$ into the differential equation we obtain

$$(x-1)y'' + 3y = (-2c_2 + 3c_0) + \sum_{k=3}^{\infty}(k-1)(k-2)c_{k-1} - k(k-1)c_k + 3c_{k-2}]x^{k-2} = 0$$

which implies $c_2 = 3c_0/2$ and

$$c_k = \frac{(k-1)(k-2)c_{k-1} + 3c_{k-2}}{k(k-1)}, \quad k = 3,4,5,\dots .$$

Choosing $c_0 = 1$ and $c_1 = 0$ we find

$$c_2 = \frac{3}{2}, \quad c_3 = \frac{1}{2}, \quad c_4 = \frac{5}{8}$$

and so on. For $c_0 = 0$ and $c_1 = 1$ we obtain

$$c_2 = 0, \quad c_3 = \frac{1}{2}, \quad c_4 = \frac{1}{4}$$

and so on. Thus, two solutions are

$$y_1 = C_1\left(1 + \frac{3}{2}x^2 + \frac{1}{2}x^3 + \frac{5}{8}x^4 + \cdots\right)$$

and

$$y_2 = C_2\left(x + \frac{1}{2}x^3 + \frac{1}{4}x^4 + \cdots\right).$$

18. Substituting $y = \sum_{n=0}^{\infty} c_n x^{n+r}$ into the differential equation we obtain

$$2xy'' + y' + y = \left(2r^2 - r\right) c_0 x^{r-1} + \sum_{k=1}^{\infty} [2(k+r)(k+r-1)c_k + (k+r)c_k + c_{k-1}]x^{k+r-1} = 0$$

which implies

$$2r^2 - r = r(2r-1) = 0$$

and

$$(k+r)(2k+2r-1)c_k + c_{k-1} = 0.$$

The indicial roots are $r = 0$ and $r = 1/2$. For $r = 0$ the recurrence relation is

$$c_k = -\frac{c_{k-1}}{k(2k-1)}, \quad k = 1, 2, 3, \ldots,$$

so

$$c_1 = -c_0, \qquad c_2 = \frac{1}{6}c_0, \qquad c_3 = -\frac{1}{90}c_0.$$

For $r = 1/2$ the recurrence relation is

$$c_k = -\frac{c_{k-1}}{k(2k+1)}, \quad k = 1, 2, 3, \ldots,$$

so

$$c_1 = -\frac{1}{3}c_0, \qquad c_2 = \frac{1}{30}c_0, \qquad c_3 = -\frac{1}{630}c_0.$$

Two linearly independent solutions are

$$y_1 = C_1 x \left(1 - x + \frac{1}{6}x^2 - \frac{1}{90}x^3 + \cdots\right)$$

and

$$y_2 = C_2 x^{1/2} \left(1 - \frac{1}{3}x + \frac{1}{30}x^2 - \frac{1}{630}x^3 + \cdots\right).$$

21. Substituting $y = \sum_{n=0}^{\infty} c_n x^{n+r}$ into the differential equation we obtain

$$xy'' - (2x-1)y' + (x-1)y = r^2 c_0 x^{r-1} + \left[\left(r^2 + 2r + 1\right) c_1 - (2r+1)c_0\right] x^r$$

$$+ \sum_{k=2}^{\infty} [(k+r)(k+r-1)c_k + (k+r)c_k - 2(k+r-1)c_{k-1} - c_{k-1} + c_{k-2}]x^{k+r-1}$$

$$= 0$$

which implies

$$r^2 = 0,$$

$$(r+1)^2 c_1 - (2r+1)c_0 = 0,$$

and

$$(k+r)^2 c_k - (2k+2r-1)c_{k-1} + c_{k-2} = 0.$$

The indicial roots are $r_1 = r_2 = 0$, so $c_1 = c_0$ and

$$c_k = \frac{(2k-1)c_{k-1} - c_{k-2}}{k^2}, \quad k = 2, 3, 4, \ldots .$$

Thus

$$c_2 = \frac{1}{2}c_0, \qquad c_3 = \frac{1}{3!}c_0, \qquad c_4 = \frac{1}{4!}c_0$$

and one solution is

$$y_1 = c_0\left(1 + x + \frac{1}{2}x^2 + \frac{1}{3!}x^3 + \frac{1}{4!}x^4 + \cdots\right) = c_0 e^x.$$

A second solution is

$$y_2 = e^x \int \frac{e^{\int(2-1/x)dx}}{e^{2x}}\, dx = e^x \int \frac{e^{2x}dx}{xe^{2x}} = e^x \int \frac{1}{x}\, dx = e^x \ln x.$$

7 Laplace Transform

3. $\mathcal{L}\{f(t)\} = \int_0^1 te^{-st}dt + \int_1^\infty e^{-st}dt = \left(-\frac{1}{s}te^{-st} - \frac{1}{s^2}e^{-st}\right)\Big|_0^1 - \frac{1}{s}e^{-st}\Big|_1^\infty$

$$= \left(-\frac{1}{s}e^{-s} - \frac{1}{s^2}e^{-s}\right) - \left(0 - \frac{1}{s^2}\right) - \frac{1}{s}(0 - e^{-s}) = \frac{1}{s^2}(1 - e^{-s}), \quad s > 0$$

6. $\mathcal{L}\{f(t)\} = \int_{\pi/2}^\infty (\cos t)e^{-st}dt = \left(-\frac{s}{s^2+1}e^{-st}\cos t + \frac{1}{s^2+1}e^{-st}\sin t\right)\Big|_{\pi/2}^\infty$

$$= 0 - \left(0 + \frac{1}{s^2+1}e^{-\pi s/2}\right) = -\frac{1}{s^2+1}e^{-\pi s/2}, \quad s > 0$$

9. $f(t) = \begin{cases} 1 - t, & 0 < t < 1 \\ 0, & t > 0 \end{cases}$

$$\mathcal{L}\{f(t)\} = \int_0^1 (1 - t)e^{-st}\,dt = \left(-\frac{1}{s}(1 - t)e^{-st} + \frac{1}{s^2}e^{-st}\right)\Big|_0^1 = \frac{1}{s^2}e^{-s} + \frac{1}{s} - \frac{1}{s^2}, \quad s > 0$$

12. $\mathcal{L}\{f(t)\} = \int_0^\infty e^{-2t-5}e^{-st}dt = e^{-5}\int_0^\infty e^{-(s+2)t}dt = -\frac{e^{-5}}{s+2}e^{-(s+2)t}\Big|_0^\infty = \frac{e^{-5}}{s+2}, \quad s > -2$

15. $\mathcal{L}\{f(t)\} = \int_0^\infty e^{-t}(\sin t)e^{-st}dt = \int_0^\infty (\sin t)e^{-(s+1)t}dt$

$$= \left(\frac{-(s+1)}{(s+1)^2+1}e^{-(s+1)t}\sin t - \frac{1}{(s+1)^2+1}e^{-(s+1)t}\cos t\right)\Big|_0^\infty$$

$$= \frac{1}{(s+1)^2+1} = \frac{1}{s^2+2s+2}, \quad s > -1$$

18. $\mathcal{L}\{f(t)\} = \int_0^\infty t(\sin t)e^{-st}dt$

$$= \left[\left(-\frac{t}{s^2+1} - \frac{2s}{(s^2+1)^2}\right)(\cos t)e^{-st} - \left(\frac{st}{s^2+1} + \frac{s^2-1}{(s^2+1)^2}\right)(\sin t)e^{-st}\right]_0^\infty$$

$$= \frac{2s}{(s^2+1)^2}, \quad s > 0$$

21. $\mathcal{L}\{4t - 10\} = \frac{4}{s^2} - \frac{10}{s}$

24. $\mathcal{L}\{-4t^2 + 16t + 9\} = -4\frac{2}{s^3} + \frac{16}{s^2} + \frac{9}{s}$

27. $\mathcal{L}\{1 + e^{4t}\} = \frac{1}{s} + \frac{1}{s-4}$

30. $\mathscr{L}\{e^{2t} - 2 + e^{-2t}\} = \dfrac{1}{s-2} - \dfrac{2}{s} + \dfrac{1}{s+2}$

33. $\mathscr{L}\{\sinh kt\} = \dfrac{k}{s^2 - k^2}$

36. $\mathscr{L}\{e^{-t}\cosh t\} = \mathscr{L}\left\{e^{-t}\dfrac{e^t + e^{-t}}{2}\right\} = \mathscr{L}\left\{\dfrac{1}{2} + \dfrac{1}{2}e^{-2t}\right\} = \dfrac{1}{2s} + \dfrac{1}{2(s+2)}$

39. $\mathscr{L}\{\cos t \cos 2t\} = \mathscr{L}\left\{\dfrac{1}{2}\cos 3t + \dfrac{1}{2}\cos t\right\} = \dfrac{1}{2}\dfrac{s}{s^2 + 9} + \dfrac{1}{2}\dfrac{s}{s^2 + 1}$

42. $\mathscr{L}\{\sin^3 t\} = \mathscr{L}\left\{\sin t\left(\dfrac{1}{2} - \dfrac{1}{2}\cos 2t\right)\right\} = \mathscr{L}\left\{\dfrac{1}{2}\sin t - \dfrac{1}{2}\left(\dfrac{1}{2}\sin 3t - \dfrac{1}{2}\sin t\right)\right\} = \dfrac{3}{4}\dfrac{1}{s^2 + 1} - \dfrac{1}{4}\dfrac{3}{s^2 + 9}$

45. $\mathscr{L}\{t^{1/2}\} = \dfrac{\Gamma(3/2)}{s^{3/2}} = \dfrac{\sqrt{\pi}}{2s^{3/2}}$

48. Since f and g are of exponential order there exist numbers c, d, M, and N such that $|f(t)| \leq Me^{ct}$ and $|g(t)| \leq Ne^{dt}$ for $t > T$. Then

$$|(fg)(t)| = |f(t)||g(t)| \leq Me^{ct}Ne^{dt} = MNe^{(c+d)t}$$

for $t > T$, and fg is of exponential order.

——————— **Exercises 7.2** ———————

3. $\mathscr{L}^{-1}\left\{\dfrac{1}{s^2} - \dfrac{48}{s^5}\right\} = \mathscr{L}^{-1}\left\{\dfrac{1}{s^2} - \dfrac{48}{24}\cdot\dfrac{4!}{s^5}\right\} = t - 2t^4$

6. $\mathscr{L}^{-1}\left\{\dfrac{(s+2)^3}{s^3}\right\} = \mathscr{L}^{-1}\left\{\dfrac{1}{s} + 4\cdot\dfrac{1}{s^2} + 2\cdot\dfrac{2}{s^3}\right\} = 1 + 4t + 2t^2$

9. $\mathscr{L}^{-1}\left\{\dfrac{1}{4s+1}\right\} = \mathscr{L}^{-1}\left\{\dfrac{1}{4}\cdot\dfrac{1}{s+1/4}\right\} = \dfrac{1}{4}e^{-t/4}$

12. $\mathscr{L}^{-1}\left\{\dfrac{10s}{s^2 + 16}\right\} = 10\cos 4t$

15. $\mathscr{L}^{-1}\left\{\dfrac{1}{s^2 - 16}\right\} = \mathscr{L}^{-1}\left\{\dfrac{1/8}{s-4} - \dfrac{1/8}{s+4}\right\} = \dfrac{1}{8}e^{4t} - \dfrac{1}{8}e^{-4t} = \dfrac{1}{4}\sinh 4t$

18. $\mathscr{L}^{-1}\left\{\dfrac{s+1}{s^2 + 2}\right\} = \mathscr{L}^{-1}\left\{\dfrac{s}{s^2 + 2} + \dfrac{1}{\sqrt{2}}\cdot\dfrac{\sqrt{2}}{s^2 + 2}\right\} = \cos\sqrt{2}\,t + \dfrac{1}{\sqrt{2}}\sin\sqrt{2}\,t$

21. $\mathscr{L}^{-1}\left\{\dfrac{s}{s^2 + 2s - 3}\right\} = \mathscr{L}^{-1}\left\{\dfrac{1}{4}\cdot\dfrac{1}{s-1} + \dfrac{3}{4}\cdot\dfrac{1}{s+3}\right\} = \dfrac{1}{4}e^t + \dfrac{3}{4}e^{-3t}$

24. $\mathscr{L}^{-1}\left\{\dfrac{s-3}{(s-\sqrt{3})(s+\sqrt{3})}\right\} = \mathscr{L}^{-1}\left\{\dfrac{s}{s^2 - 3} - \sqrt{3}\cdot\dfrac{\sqrt{3}}{s^2 - 3}\right\} = \cosh\sqrt{3}\,t - \sqrt{3}\sinh\sqrt{3}\,t$

27. $\mathscr{L}^{-1}\left\{\dfrac{2s+4}{(s-2)(s^2 + 4s + 3)}\right\} = \mathscr{L}^{-1}\left\{\dfrac{8}{15}\cdot\dfrac{1}{s-2} - \dfrac{1}{3}\cdot\dfrac{1}{s+1} - \dfrac{1}{5}\cdot\dfrac{1}{s+3}\right\} = \dfrac{8}{15}e^{2t} - \dfrac{1}{3}e^{-t} - \dfrac{1}{5}e^{-3t}$

30. $\mathcal{L}^{-1}\left\{\dfrac{s-1}{s^2(s^2+1)}\right\} = \mathcal{L}^{-1}\left\{\dfrac{1}{s} - \dfrac{1}{s^2} - \dfrac{s}{s^2+1} + \dfrac{1}{s^2+1}\right\} = 1 - t - \cos t + \sin t$

33. $\mathcal{L}^{-1}\left\{\dfrac{1}{(s^2+1)(s^2+4)}\right\} = \mathcal{L}^{-1}\left\{\dfrac{1}{3}\cdot\dfrac{1}{s^2+1} - \dfrac{1}{6}\cdot\dfrac{2}{s^2+4}\right\} = \dfrac{1}{3}\sin t - \dfrac{1}{6}\sin 2t$

36. $\mathcal{L}\{f(t)\} = \displaystyle\int_0^\infty e^{(3-s)t}dt = \dfrac{1}{s-3} \quad \text{for } s > 3$

Exercises 7.3

3. $\mathcal{L}\left\{t^3 e^{-2t}\right\} = \dfrac{3!}{(s+2)^4}$

6. $\mathcal{L}\left\{e^{-2t}\cos 4t\right\} = \dfrac{s+2}{(s+2)^2+16}$

9. $\mathcal{L}\left\{t\left(e^t + e^{2t}\right)^2\right\} = \mathcal{L}\left\{te^{2t} + 2te^{3t} + te^{4t}\right\} = \dfrac{1}{(s-2)^2} + \dfrac{2}{(s-3)^2} + \dfrac{1}{(s-4)^2}$

12. $\mathcal{L}\left\{e^t\cos^2 3t\right\} = \mathcal{L}\left\{\dfrac{1}{2}e^t + \dfrac{1}{2}e^t\cos 6t\right\} = \dfrac{1}{2}\dfrac{1}{s-1} + \dfrac{1}{2}\dfrac{s-1}{(s-1)^2+36}$

15. $\mathcal{L}^{-1}\left\{\dfrac{1}{s^2-6s+10}\right\} = \mathcal{L}^{-1}\left\{\dfrac{1}{(s-3)^2+1^2}\right\} = e^{3t}\sin t$

18. $\mathcal{L}^{-1}\left\{\dfrac{2s+5}{s^2+6s+34}\right\} = \mathcal{L}^{-1}\left\{2\dfrac{(s+3)}{(s+3)^2+5^2} - \dfrac{1}{5}\dfrac{5}{(s+3)^2+5^2}\right\} = 2e^{-3t}\cos 5t - \dfrac{1}{5}e^{-3t}\sin 5t$

21. $\mathcal{L}^{-1}\left\{\dfrac{2s-1}{s^2(s+1)^3}\right\} = \mathcal{L}^{-1}\left\{\dfrac{5}{s} - \dfrac{1}{s^2} - \dfrac{5}{s+1} - \dfrac{4}{(s+1)^2} - \dfrac{3}{2}\dfrac{2}{(s+1)^3}\right\} = 5 - t - 5e^{-t} - 4te^{-t} - \dfrac{3}{2}t^2e^{-t}$

24. $\mathcal{L}\{e^{2-t}\mathcal{U}(t-2)\} = \mathcal{L}\left\{e^{-(t-2)}\mathcal{U}(t-2)\right\} = \dfrac{e^{-2s}}{s+1}$

27. $\mathcal{L}\{\cos 2t\,\mathcal{U}(t-\pi)\} = \mathcal{L}\left\{\cos 2(t-\pi)\,\mathcal{U}(t-\pi)\right\} = \dfrac{se^{-\pi s}}{s^2+4}$

30. $\mathcal{L}\left\{te^{t-5}\mathcal{U}(t-5)\right\} = \mathcal{L}\left\{(t-5)e^{t-5}\mathcal{U}(t-5) + 5e^{t-5}\mathcal{U}(t-5)\right\} = \dfrac{e^{-5s}}{(s-1)^2} + \dfrac{5e^{-5s}}{s-1}$

33. $\mathcal{L}^{-1}\left\{\dfrac{e^{-\pi s}}{s^2+1}\right\} = \sin(t-\pi)\,\mathcal{U}(t-\pi)$

36. $\mathcal{L}^{-1}\left\{\dfrac{e^{-2s}}{s^2(s-1)}\right\} = \mathcal{L}^{-1}\left\{-\dfrac{e^{-2s}}{s} - \dfrac{e^{-2s}}{s^2} + \dfrac{e^{-2s}}{s-1}\right\} = -\mathcal{U}(t-2) - (t-2)\mathcal{U}(t-2) + e^{t-2}\mathcal{U}(t-2)$

39. $\mathcal{L}\{t^2\sinh t\} = \dfrac{d^2}{ds^2}\left(\dfrac{1}{s^2-1}\right) = \dfrac{6s^2+2}{(s^2-1)^3}$

42. $\mathcal{L}\left\{te^{-3t}\cos 3t\right\} = -\dfrac{d}{ds}\left(\dfrac{s+3}{(s+3)^2+9}\right) = \dfrac{(s+3)^2-9}{[(s+3)^2+9]^2}$

45. (c)

48. (b)

51. $\mathcal{L}\{2 - 4\,\mathcal{U}(t-3)\} = \dfrac{2}{s} - \dfrac{4}{s}e^{-3s}$

54. $\mathcal{L}\left\{\sin t\,\mathcal{U}\left(t - \dfrac{3\pi}{2}\right)\right\} = \mathcal{L}\left\{-\cos\left(t - \dfrac{3\pi}{2}\right)\mathcal{U}\left(t - \dfrac{3\pi}{2}\right)\right\} = -\dfrac{se^{-3\pi s/2}}{s^2 + 1}$

57. $\mathcal{L}\{f(t)\} = \mathcal{L}\{\mathcal{U}(t-a) - \mathcal{U}(t-b)\} = \dfrac{e^{-as}}{s} - \dfrac{e^{-bs}}{s}$

60. $\mathcal{L}^{-1}\left\{\dfrac{2}{s} - \dfrac{3e^{-s}}{s^2} + \dfrac{5e^{-2s}}{s^2}\right\} = 2 - 3(t-1)\,\mathcal{U}(t-1) + 5(t-2)\,\mathcal{U}(t-2)$

$$= \begin{cases} 2, & 0 \leq t < 1 \\ -3t + 5, & 1 \leq t < 2 \\ 2t - 5, & t \geq 2 \end{cases}$$

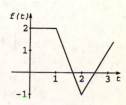

63. $f(t) = -\dfrac{1}{t}\mathcal{L}^{-1}\left\{\dfrac{d}{ds}\left(\dfrac{\pi}{2} - \tan^{-1}\dfrac{s}{2}\right)\right\} = -\dfrac{1}{t}\mathcal{L}^{-1}\left\{-\dfrac{2}{s^2 + 2^2}\right\} = \dfrac{\sin 2t}{t}$

Exercises 7.4

3. $\mathcal{L}\{y'' + 3y'\} = \mathcal{L}\{y''\} + 3\mathcal{L}\{y'\} = s^2 Y(s) - sy(0) - y'(0) + 3[sY(s) - y(0)] = (s^2 + 3s)Y(s) - s - 2$

6. We solve $\mathcal{L}\{y'' + y\} = \mathcal{L}\{1\} = 1/s$.

$$s^2 Y(s) - sy(0) - y'(0) + Y(s) = \dfrac{1}{s}$$

$$(s^2 + 1)Y(s) - 2s - 3 = \dfrac{1}{s}$$

$$Y(s) = \dfrac{1}{s(s^2 + 1)} + \dfrac{2s + 3}{s^2 + 1}$$

9. $\mathcal{L}\left\{\displaystyle\int_0^t e^{-\tau}\cos\tau\,d\tau\right\} = \dfrac{1}{s}\mathcal{L}\{e^{-t}\cos t\} = \dfrac{1}{s}\dfrac{s+1}{(s+1)^2 + 1} = \dfrac{s+1}{s(s^2 + 2s + 2)}$

12. $\mathcal{L}\left\{\displaystyle\int_0^t \sin\tau\cos(t-\tau)\,d\tau\right\} = \mathcal{L}\{\sin t\}\mathcal{L}\{\cos t\} = \dfrac{s}{(s^2 + 1)^2}$

15. $\mathcal{L}\{1 * t^3\} = \dfrac{1}{s}\dfrac{3!}{s^4} = \dfrac{6}{s^5}$

18. $\mathcal{L}\{t^2 * te^t\} = \dfrac{2}{s^3(s-1)^2}$

21. $\mathcal{L}^{-1}\left\{\dfrac{1}{s+5}F(s)\right\} = e^{-5t} * f(t) = \displaystyle\int_0^t f(\tau)e^{-5(t-\tau)}\,d\tau$

24. $\mathcal{L}^{-1}\left\{\dfrac{1}{s(s^2+1)}\right\} = 1 * \sin t = \displaystyle\int_0^t \sin(t-\tau)\,d\tau = \cos(t-\tau)\,\Big|_0^t = 1 - \cos t$

27. $\mathcal{L}^{-1}\left\{\dfrac{s}{(s^2+4)^2}\right\} = \cos 2t * \dfrac{1}{2}\sin 2t = \dfrac{1}{2}\displaystyle\int_0^t \cos 2\tau \sin 2(t-\tau)\,d\tau$

$$= \dfrac{1}{2}\int_0^t \cos 2\tau(\sin 2t \cos 2\tau - \cos 2t \sin 2\tau)\,d\tau = \dfrac{1}{2}\left[\sin 2t \int_0^t \cos^2 2\tau\,d\tau - \cos 2t \int_0^t \dfrac{1}{2}\sin 4\tau\,d\tau\right]$$

$$= \dfrac{1}{2}\sin 2t\left[\dfrac{1}{2}\tau + \dfrac{1}{8}\sin 4\tau\right]_0^t - \dfrac{1}{4}\cos 2t\left[-\dfrac{1}{4}\cos 4\tau\right]_0^t$$

$$= \dfrac{1}{2}\sin 2t\left(\dfrac{1}{2}t + \dfrac{1}{8}\sin 4t\right) + \dfrac{1}{16}\cos 2t(\cos 4t - 1)$$

$$= \dfrac{1}{4}t\sin 2t + \dfrac{1}{16}\sin 2t \sin 4t + \dfrac{1}{16}\cos 2t \cos 4t - \dfrac{1}{16}\cos 2t$$

$$= \dfrac{1}{4}t\sin 2t + \dfrac{1}{16}\left[\sin 2t(2\sin 2t \cos 2t) + \cos 2t\left(\cos^2 2t - \sin^2 2t\right) - \cos 2t\right]$$

$$= \dfrac{1}{4}t\sin 2t + \dfrac{1}{16}\cos 2t\left[2\sin^2 2t + \cos^2 2t - \sin^2 2t - 1\right] = \dfrac{1}{4}t\sin 2t$$

30. $f * (g+h) = \displaystyle\int_0^t f(\tau)[g(t-\tau) + h(t-\tau)]\,d\tau = \int_0^t f(\tau)g(t-\tau)\,d\tau + \int_0^t f(\tau)h(t-\tau)\,d\tau$

$$= \int_0^t f(\tau)[g(t-\tau) + h(t-\tau)]\,d\tau = f * g + f * h$$

33. $\mathcal{L}\{f(t)\} = \dfrac{1}{1-e^{-bs}}\displaystyle\int_0^b \dfrac{a}{b}te^{-st}dt = \dfrac{a}{s}\left(\dfrac{1}{bs} - \dfrac{1}{e^{bs}-1}\right)$

36. $\mathcal{L}\{f(t)\} = \dfrac{1}{1-e^{-2\pi s}}\displaystyle\int_0^\pi e^{-st}\sin t\,dt = \dfrac{1}{s^2+1}\cdot\dfrac{1}{1-e^{-\pi s}}$

Exercises 7.5

3. The Laplace transform of the differential equation is

$$s\,\mathcal{L}\{y\} - y(0) + 4\,\mathcal{L}\{y\} = \dfrac{1}{s+4}.$$

Solving for $\mathcal{L}\{y\}$ we obtain $\mathcal{L}\{y\} = \dfrac{1}{(s+4)^2} + \dfrac{2}{s+4}.$

Thus, $\qquad\qquad\qquad\qquad y = te^{-4t} + 2e^{-4t}.$

6. The Laplace transform of the differential equation is

$$s^2\,\mathcal{L}\{y\} - sy(0) - y'(0) - 6\left[s\,\mathcal{L}\{y\} - y(0)\right] + 13\,\mathcal{L}\{y\} = 0.$$

Solving for $\mathscr{L}\{y\}$ we obtain

$$\mathscr{L}\{y\} = -\frac{3}{s^2 - 6s + 13} = -\frac{3}{2}\frac{2}{(s-3)^2 + 2^2}.$$

Thus,

$$y = -\frac{3}{2}e^{3t}\sin 2t.$$

9. The Laplace transform of the differential equation is

$$s^2\mathscr{L}\{y\} - sy(0) - y'(0) - 4\left[s\mathscr{L}\{y\} - y(0)\right] + 4\mathscr{L}\{y\} = \frac{6}{(s-2)^4}.$$

Solving for $\mathscr{L}\{y\}$ we obtain $\quad \mathscr{L}\{y\} = \dfrac{1}{20}\dfrac{5!}{(s-2)^6}$. Thus, $y = \dfrac{1}{20}t^5 e^{2t}$.

12. The Laplace transform of the differential equation is

$$s^2\mathscr{L}\{y\} - sy(0) - y'(0) + 16\,\mathscr{L}\{y\} = \frac{1}{s}.$$

Solving for $\mathscr{L}\{y\}$ we obtain

$$\mathscr{L}\{y\} = \frac{s^2 + 2s + 1}{s(s^2 + 16)} = \frac{1}{16}\frac{1}{s} + \frac{15}{16}\frac{s}{s^2 + 4^2} + \frac{1}{2}\frac{4}{s^2 + 4^2}.$$

Thus,

$$y = \frac{1}{16} + \frac{15}{16}\cos 4t + \frac{1}{2}\sin 4t.$$

15. The Laplace transform of the differential equation is

$$2\left[s^3\mathscr{L}\{y\} - s^2(0) - sy'(0) - y''(0)\right] + 3[s^2\mathscr{L}\{y\} - sy(0) - y'(0)] - 3[s\mathscr{L}\{y\} - y(0)] - 2\mathscr{L}\{y\} = \frac{1}{s+1}.$$

Solving for $\mathscr{L}\{y\}$ we obtain

$$\mathscr{L}\{y\} = \frac{2s+3}{(s+1)(s-1)(2s+1)(s+2)} = \frac{1}{2}\frac{1}{s+1} + \frac{5}{18}\frac{1}{s-1} - \frac{8}{9}\frac{1}{s+1/2} + \frac{1}{9}\frac{1}{s+2}.$$

Thus,

$$y = \frac{1}{2}e^{-t} + \frac{5}{18}e^{t} - \frac{8}{9}e^{-t/2} + \frac{1}{9}e^{-2t}.$$

18. The Laplace transform of the differential equation is

$$s^4\mathscr{L}\{y\} - s^3y(0) - s^2y'(0) - sy''(0) - y'''(0) - \mathscr{L}\{y\} = \frac{1}{s^2}.$$

Solving for $\mathscr{L}\{y\}$ we obtain

$$\mathscr{L}\{y\} = \frac{1}{s^2(s^4 - 1)} = -\frac{1}{s^2} + \frac{1}{4}\frac{1}{s-1} - \frac{1}{4}\frac{1}{s+1} + \frac{1}{2}\frac{1}{s^2 + 1}.$$

Thus,

$$y = -t + \frac{1}{4}e^{t} - \frac{1}{4}e^{-t} + \frac{1}{2}\sin t.$$

21. The Laplace transform of the differential equation is

$$s\,\mathcal{L}\{y\} - y(0) + 2\,\mathcal{L}\{y\} = \frac{1}{s^2} - e^{-s}\frac{s+1}{s^2}.$$

Solving for $\mathcal{L}\{y\}$ we obtain

$$\mathcal{L}\{y\} = \frac{1}{s^2(s+2)} - e^{-s}\frac{s+1}{s^2(s+1)} = -\frac{1}{4}\frac{1}{s} + \frac{1}{2}\frac{1}{s^2} + \frac{1}{4}\frac{1}{s+2} - e^{-s}\left[\frac{1}{4}\frac{1}{s} + \frac{1}{2}\frac{1}{s^2} - \frac{1}{4}\frac{1}{s+2}\right].$$

Thus,

$$y = -\frac{1}{4} + \frac{1}{2}t + \frac{1}{4}e^{-2t} - \left[\frac{1}{4} + \frac{1}{2}(t-1) - \frac{1}{4}e^{-2(t-1)}\right]\mathcal{U}(t-1).$$

24. The Laplace transform of the differential equation is

$$s^2\,\mathcal{L}\{y\} - sy(0) - y'(0) - 5\left[s\,\mathcal{L}\{y\} - y(0)\right] + 6\,\mathcal{L}\{y\} = \frac{e^{-s}}{s}.$$

Solving for $\mathcal{L}\{y\}$ we obtain

$$\mathcal{L}\{y\} = e^{-s}\frac{1}{s(s-2)(s-3)} + \frac{1}{(s-2)(s-3)}$$

$$= e^{-s}\left[\frac{1}{6}\frac{1}{s} - \frac{1}{2}\frac{1}{s-2} + \frac{1}{3}\frac{1}{s-3}\right] - \frac{1}{s-2} + \frac{1}{s-3}.$$

Thus,

$$y = \left[\frac{1}{6} - \frac{1}{2}e^{2(t-1)} + \frac{1}{3}e^{3(t-1)}\right]\mathcal{U}(t-1) + e^{3t} - e^{2t}.$$

27. Taking the Laplace transform of both sides of the differential equation and letting $c = y(0)$ we obtain

$$\mathcal{L}\{y''\} + \mathcal{L}\{2y'\} + \mathcal{L}\{y\} = 0$$

$$s^2\mathcal{L}\{y\} - sy(0) - y'(0) + 2s\,\mathcal{L}\{y\} - 2y(0) + \mathcal{L}\{y\} = 0$$

$$s^2\mathcal{L}\{y\} - cs - 2 + 2s\,\mathcal{L}\{y\} - 2c + \mathcal{L}\{y\} = 0$$

$$\left(s^2 + 2s + 1\right)\mathcal{L}\{y\} = cs + 2c + 2$$

$$\mathcal{L}\{y\} = \frac{cs}{(s+1)^2} + \frac{2c+2}{(s+1)^2}$$

$$= c\frac{s+1-1}{(s+1)^2} + \frac{2c+2}{(s+1)^2}$$

$$= \frac{c}{s+1} + \frac{c+2}{(s+1)^2}.$$

Therefore,

$$y(t) = c\,\mathcal{L}^{-1}\left\{\frac{1}{s+1}\right\} + (c+2)\,\mathcal{L}^{-1}\left\{\frac{1}{(s+1)^2}\right\} = ce^{-t} + (c+2)te^{-t}.$$

To find c we let $y(1) = 2$. Then $2 = ce^{-1} + (c+2)e^{-1} = 2(c+1)e^{-1}$ and $c = e - 1$. Thus,

$$y(t) = (e - 1)e^{-t} + (e + 1)te^{-t}.$$

30. The Laplace transform of the given equation is

$$\mathscr{L}\{f\} = \mathscr{L}\{2t\} - 4\mathscr{L}\{\sin t\}\mathscr{L}\{f\}.$$

Solving for $\mathscr{L}\{f\}$ we obtain

$$\mathscr{L}\{f\} = \frac{2s^2 + 2}{s^2(s^2 + 5)} = \frac{2}{5}\frac{1}{s^2} + \frac{8}{5\sqrt{5}}\frac{\sqrt{5}}{s^2 + 5}.$$

Thus,

$$f(t) = \frac{2}{5}t + \frac{8}{5\sqrt{5}}\sin\sqrt{5}\,t.$$

33. The Laplace transform of the given equation is

$$\mathscr{L}\{f\} + \mathscr{L}\{1\}\mathscr{L}\{f\} = \mathscr{L}\{1\}.$$

Solving for $\mathscr{L}\{f\}$ we obtain $\mathscr{L}\{f\} = \dfrac{1}{s+1}$. Thus, $f(t) = e^{-t}$.

36. The Laplace transform of the given equation is

$$\mathscr{L}\{t\} - 2\mathscr{L}\{f\} = \mathscr{L}\{e^t - e^{-t}\}\mathscr{L}\{f\}.$$

Solving for $\mathscr{L}\{f\}$ we obtain

$$\mathscr{L}\{f\} = \frac{s^2 - 1}{2s^4} = \frac{1}{2}\frac{1}{s^2} - \frac{1}{12}\frac{3!}{s^4}.$$

Thus,

$$f(t) = \frac{1}{2}t - \frac{1}{12}t^3.$$

39. From equation (3) in the text the differential equation is

$$0.005\frac{di}{dt} + i + 50\int_0^t i(\tau)\,d\tau = 100[1 - \mathscr{U}(t - 1)], \quad i(0) = 0.$$

The Laplace transform of this equation is

$$0.005[s\,\mathscr{L}\{i\} - i(0)] + \mathscr{L}\{i\} + 50\frac{1}{s}\mathscr{L}\{i\} = 100\left[\frac{1}{s} - \frac{1}{s}e^{-s}\right].$$

Solving for $\mathscr{L}\{i\}$ we obtain

$$\mathscr{L}\{i\} = \frac{20{,}000}{(s + 100)^2}(1 - e^{-s}]).$$

Thus,

$$i(t) = 20{,}000te^{-100t} - 20{,}000(t - 1)e^{-100(t-1)}\mathscr{U}(t - 1).$$

42. The differential equation is

$$10\frac{dq}{dt} + 10q = 30e^t - 30e^t\,\mathcal{U}(t - 1.5).$$

The Laplace transform of this equation is

$$s\,\mathcal{L}\{q\} - q_0 + \mathcal{L}\{q\} = \frac{3}{s-1} - \frac{3e^{1.5}}{s-1.5}e^{-1.5s}.$$

Solving for $\mathcal{L}\{q\}$ we obtain

$$\mathcal{L}\{q\} = \left(q_0 - \frac{3}{2}\right)\cdot\frac{1}{s+1} + \frac{3}{2}\cdot\frac{1}{s-1}3e^{1.5}\left(\frac{-2/5}{s+1} + \frac{2/5}{s-1.5}\right)e^{-1.55}.$$

Thus,

$$q(t) = \left(q_0 - \frac{3}{2}\right)e^{-t} + \frac{3}{2}e^t + \frac{6}{5}e^{1.5}\left(e^{-(t-1.5)} - e^{1.5(t-1.5)}\right)\mathcal{U}(t - 1.5).$$

45. The differential equation is

$$\frac{di}{dt} + 10i = \sin t + \cos\left(t - \frac{3\pi}{2}\right)\mathcal{U}\left(t - \frac{3\pi}{2}\right), \quad i(0) = 0.$$

The Laplace transform of this equation is

$$s\,\mathcal{L}\{i\} + 10\mathcal{L}\{i\} = \frac{1}{s^2 + 1} + \frac{se^{-3\pi s/2}}{s^2 + 1}.$$

Solving for $\mathcal{L}\{i\}$ we obtain

$$\mathcal{L}\{i\} = \frac{1}{(s^2 + 1)(s + 10)} + \frac{s}{(s^2 + 1)(s + 10)}e^{-3\pi s/2}$$

$$= \frac{1}{101}\left(\frac{1}{s+10} - \frac{s}{s^2+1} + \frac{10}{s^2+1}\right) + \frac{1}{101}\left(\frac{-10}{s+10} + \frac{10s}{s^2+1} + \frac{1}{s^2+1}\right)e^{-3\pi s/2}.$$

Thus,

$$i(t) = \frac{1}{101}\left(e^{-10t} - \cos t + 10\sin t\right)$$

$$+ \frac{1}{101}\left(-10e^{-10(t-3\pi/2)} + 10\cos\left(t - \frac{3\pi}{2}\right) + \sin\left(t - \frac{3\pi}{2}\right)\right]\mathcal{U}\left(t - \frac{3\pi}{2}\right).$$

48. The differential equation is

$$\frac{d^2q}{dt^2} + 20\frac{dq}{dt} + 200q = 150, \quad q(0) = q'(0) = 0.$$

The Laplace transform of this equation is

$$s^2\mathcal{L}\{q\} + 20s\mathcal{L}\{q\} + 200\mathcal{L}\{q\} = \frac{150}{s}.$$

Solving for $\mathcal{L}\{q\}$ we obtain

$$\mathcal{L}\{q\} = \frac{150}{s(s^2 + 20s + 200)} = \frac{3}{4}\frac{1}{s} - \frac{3}{4}\frac{s+10}{(s+10)^2 + 10^2} - \frac{3}{4}\frac{10}{(s+10)^2 + 10^2}.$$

Thus,

$$q(t) = \frac{3}{4} - \frac{3}{4}e^{-10t}\cos 10t - \frac{3}{4}e^{-10t}\sin 10t$$

and

$$i(t) = q'(t) = 15e^{-10t}\sin 10t.$$

If $E(t) = 150 - 150\,\mathcal{U}(t-2)$, then

$$\mathcal{L}\{q\} = \frac{150}{s(s^2 + 20s + 200)}\left(1 - e^{-2s}\right)$$

$$q(t) = \frac{3}{4} - \frac{3}{4}e^{-10t}\cos 10t - \frac{3}{4}e^{-10t}\sin 10t - \left[\frac{3}{4} - \frac{3}{4}e^{-10(t-2)}\cos 10(t-2)\right.$$

$$\left. - \frac{3}{4}e^{-10(t-2)}\sin 10(t-2)\right]\mathcal{U}(t-2).$$

51. The differential equation is

$$\frac{d^2q}{dt^2} + \frac{1}{LC}q = \frac{E_0}{L}e^{-kt}, \quad q(0) = q'(0) = 0.$$

The Laplace transform of this equation is

$$s^2\mathcal{L}\{q\} + \frac{1}{LC}\mathcal{L}\{q\} = \frac{E_0}{L}\frac{1}{s+k}.$$

Solving for $\mathcal{L}\{q\}$ we obtain

$$\mathcal{L}\{q\} = \frac{E_0}{L}\frac{1}{(s+k)(s^2 + 1/LC)} = \frac{E_0}{L}\left(\frac{1/(k^2 + 1/LC)}{s+k} - \frac{s/(k^2 + 1/LC)}{s^2 + 1/LC} + \frac{k/(k^2 + 1/LC)}{s^2 + 1/LC}\right).$$

Thus,

$$q(t) = \frac{E_0}{L(k^2 + 1/LC)}\left[e^{-kt} - \cos\left(t/\sqrt{LC}\right) + k\sqrt{LC}\sin\left(t/\sqrt{LC}\right)\right].$$

54. Recall from Chapter 5 that $mx'' = -kx + f(t)$. Now $m = W/g = 16/32 = 1/2$ slug, and $k = 4.5$, so the differential equation is

$$\frac{1}{2}x'' + 4.5x = 4\sin 3t + 2\cos 3t \quad \text{or} \quad x'' + 9x = 8\sin 3t + 4\cos 3t.$$

The initial conditions are $x(0) = x'(0) = 0$. The Laplace transform of the differential equation is

$$s^2\mathcal{L}\{x\} + 9\mathcal{L}\{x\} = \frac{24}{s^2 + 9} + \frac{4s}{s^2 + 9}.$$

Solving for $\mathscr{L}\{x\}$ we obtain

$$\mathscr{L}\{x\} = \frac{4s+24}{(s^2+9)^2} = \frac{2}{3}\frac{2(3)s}{(s^2+9)^2} + \frac{12}{27}\frac{2(3)^3}{(s^2+9)^2}.$$

Thus,

$$x(t) = \frac{2}{3}t\sin 3t + \frac{4}{9}(\sin 3t - 3t\cos 3t) = \frac{2}{3}t\sin 3t + \frac{4}{9}\sin 3t - \frac{4}{3}t\cos 3t.$$

57. The differential equation is

$$EI\frac{d^4y}{dx^4} = w_0[1 - \mathscr{U}(x - L/2)].$$

Taking the Laplace transform of both sides and using $y(0) = y'(0) = 0$ we obtain

$$s^4\mathscr{L}\{y\} - sy''(0) - y'''(0) = \frac{w_0}{EI}\frac{1}{s}\left(1 - e^{-Ls/2}\right).$$

Letting $y''(0) = c_1$ and $y'''(0) = c_2$ we have

$$\mathscr{L}\{y\} = \frac{c_1}{s^3} + \frac{c_2}{s^4} + \frac{w_0}{EI}\frac{1}{s^5}\left(1 - e^{-Ls/2}\right)$$

so that

$$y(x) = \frac{1}{2}c_1 x^2 + \frac{1}{6}c_2 x^3 + \frac{1}{24}\frac{w_0}{EI}\left[x^4 - \left(x - \frac{L}{2}\right)^4\mathscr{U}\left(x - \frac{L}{2}\right)\right].$$

To find c_1 and c_2 we compute

$$y''(x) = c_1 + c_2 x + \frac{1}{2}\frac{w_0}{EI}\left[x^2 - \left(x - \frac{L}{2}\right)^2\mathscr{U}\left(x - \frac{L}{2}\right)\right]$$

and

$$y'''(x) = c_2 + \frac{w_0}{EI}\left[x - \left(x - \frac{L}{2}\right)\mathscr{U}\left(x - \frac{L}{2}\right)\right].$$

Then $y''(L) = y'''(L) = 0$ yields the system

$$c_1 + c_2 L + \frac{1}{2}\frac{w_0}{EI}\left[L^2 - \left(\frac{L}{2}\right)^2\right] = c_1 + c_2 L + \frac{3}{8}\frac{w_0 L^2}{EI} = 0$$

$$c_2 + \frac{w_0}{EI}\left(\frac{L}{2}\right) = c_2 + \frac{1}{2}\frac{w_0 L}{EI} = 0.$$

Solving for c_1 and c_2 we obtain $c_1 = \frac{1}{8}w_0 L^2/EI$ and $c_2 = -\frac{1}{2}w_0 L/EI$. Thus,

$$y(x) = \frac{w_0}{EI}\left(\frac{1}{16}L^2 x^2 - \frac{1}{12}Lx^3 + \frac{1}{24}x^4 - \frac{1}{24}\left(x - \frac{L}{2}\right)^4\mathscr{U}\left(x - \frac{L}{2}\right)\right).$$

60. The Laplace transform of the differential equation is

$$-\frac{d}{ds}\left[s^2\mathscr{L}\{y\} - y'(0)\right] - 2\frac{d}{ds}[s\mathscr{L}\{y\}] + 2\mathscr{L}\{y\} = 0.$$

Then

$$-s^2\left(\frac{d}{ds}\mathscr{L}\{y\}\right) - 2s\mathscr{L}\{y\} - 2s\left(\frac{d}{ds}\mathscr{L}\{y\}\right) - 2\mathscr{L}\{y\} + 2\mathscr{L}\{y\} = 0$$

and

$$\frac{d}{ds}\mathscr{L}\{y\} + \frac{2}{s+2}\mathscr{L}\{y\} = 0.$$

This is a separable differential equation so

$$\frac{d\mathscr{L}\{y\}}{\mathscr{L}\{y\}} = -\frac{2\,ds}{s+2} \implies \ln\mathscr{L}\{y\} = -2\ln(s+2) + c \implies \mathscr{L}\{y\} = c_1 e^{-2\ln(s+2)} = c_1(s+2)^{-1}$$

and $y(t) = c_1 t e^{-2t}$.

Exercises 7.6

3. The Laplace transform of the differential equation is

$$\mathscr{L}\{y\} = \frac{1}{s^2+1}\left(1 + e^{-2\pi s}\right)$$

so that

$$y = \sin t + \sin t\,\mathscr{U}(t - 2\pi).$$

6. The Laplace transform of the differential equation is

$$\mathscr{L}\{y\} = \frac{s}{s^2+1} + \frac{1}{s^2+1}(e^{-2\pi s} + e^{-4\pi s})$$

so that

$$y = \cos t + \sin t[\mathscr{U}(t - 2\pi) + \mathscr{U}(t - 4\pi)].$$

9. The Laplace transform of the differential equation is

$$\mathscr{L}\{y\} = \frac{1}{(s+2)^2+1}e^{-2\pi s}$$

so that

$$y = e^{-2(t-2\pi)}\sin t\,\mathscr{U}(t - 2\pi).$$

12. The Laplace transform of the differential equation is

$$\mathscr{L}\{y\} = \frac{1}{(s-1)^2(s-6)} + \frac{e^{-2s} + e^{-4s}}{(s-1)(s-6)}$$

$$= -\frac{1}{25}\frac{1}{s-1} - \frac{1}{5}\frac{1}{(s-1)^2} + \frac{1}{25}\frac{1}{s-6} + \left[-\frac{1}{5}\frac{1}{s-1} + \frac{1}{5}\frac{1}{s-6}\right](e^{-2s} + e^{-4s})$$

so that

$$y = -\frac{1}{25}e^t - \frac{1}{5}te^t + \frac{1}{25}e^{6t} + \left[-\frac{1}{5}e^{t-2} + \frac{1}{5}e^{6(t-2)}\right]\mathcal{U}(t-2)$$

$$+ \left[-\frac{1}{5}e^{t-4} + \frac{1}{5}e^{6(t-4)}\right]\mathcal{U}(t-4).$$

15. Letting $f(t) = e^{-st}$, we have by (7)

$$\mathcal{L}\{\delta(t - t_0)\} = \int_0^\infty e^{-st}\delta(t - t_0)\,dt = f(t_0) = e^{-st_0}.$$

18. The Laplace transform of the differential equation is

$$\mathcal{L}\{y\} = \frac{1}{w}\frac{s}{s^2 + w^2}$$

so that

$$y = \frac{1}{w}\sin wt.$$

Note that $y'(0) = 1$.

Chapter 7 Review Exercises

3. False; consider $f(t) = t^{-1/2}$.

6. False; consider $f(t) = 1$ and $g(t) = 1$.

9. $\mathcal{L}\{\sin 2t\} = \dfrac{2}{s^2 + 4}$

12. $\mathcal{L}\{\sin 2t\,\mathcal{U}(t - \pi)\} = \mathcal{L}\{\sin 2(t - \pi)\mathcal{U}(t - \pi)\} = \dfrac{2}{s^2 + 4}e^{-\pi s}$

15. $\mathcal{L}^{-1}\left\{\dfrac{1}{(s-5)^3}\right\} = \mathcal{L}^{-1}\left\{\dfrac{1}{2}\dfrac{2}{(s-5)^3}\right\} = \dfrac{1}{2}t^2 e^{5t}$

18. $\mathcal{L}^{-1}\left\{\dfrac{1}{s^2}e^{-5s}\right\} = (t - 5)\mathcal{U}(t - 5)$

21. $\mathcal{L}\{e^{-5t}\}$ exists for $s > -5$.

24. $1 * 1 = \displaystyle\int_0^t d\tau = t$

27. **(a)** $f(t) = 2 - 2\mathcal{U}(t - 2) + [(t - 2) + 2]\mathcal{U}(t - 2) = 2 + (t - 2)\mathcal{U}(t - 2)$

(b) $\mathcal{L}\{f(t)\} = \dfrac{2}{s} + \dfrac{1}{s^2}e^{-2s}$

(c) $\mathcal{L}\{e^t f(t)\} = \dfrac{2}{s - 1} + \dfrac{1}{(s - 1)^2}e^{-2(s-1)}$

30. Taking the Laplace transform of the differential equation we obtain

$$\mathscr{L}\{y\} = \frac{1}{(s-1)^2(s^2-8s+20)}$$

$$= \frac{6}{169}\frac{1}{s-1} + \frac{1}{13}\frac{1}{(s-1)^2} - \frac{6}{169}\frac{s-4}{(s-4)^2+2^2} + \frac{5}{338}\frac{2}{(s-4)^2+2^2}$$

so that

$$y = \frac{6}{169}e^t + \frac{1}{13}te^t - \frac{6}{169}e^{4t}\cos 2t + \frac{5}{338}e^{4t}\sin 2t.$$

33. Taking the Laplace transform of the differential equation we obtain

$$\mathscr{L}\{y\} = \frac{s^3+2}{s^3(s-5)} - \frac{2+2s+s^2}{s^3(s-5)}e^{-s}$$

$$= -\frac{2}{125}\frac{1}{s} - \frac{2}{25}\frac{1}{s^2} - \frac{1}{5}\frac{2}{s^3} + \frac{127}{125}\frac{1}{s-5} - \left[-\frac{37}{125}\frac{1}{s} - \frac{12}{25}\frac{1}{s^2} - \frac{1}{5}\frac{2}{s^3} + \frac{37}{125}\frac{1}{s-5}\right]e^{-s}$$

so that

$$y = -\frac{2}{125} - \frac{2}{25}t - \frac{1}{5}t^2 + \frac{127}{125}e^{5t} - \left[-\frac{37}{125} - \frac{12}{25}(t-1) - \frac{1}{5}(t-1)^2 + \frac{37}{125}e^{5(t-1)}\right]\mathcal{U}(t-1).$$

36. Taking the Laplace transform of the integral equation we obtain

$$(\mathscr{L}\{f\})^2 = 6\cdot\frac{6}{s^4} \quad \text{or} \quad \mathscr{L}\{f\} = \pm 6\cdot\frac{1}{s^2}$$

so that $f(t) = \pm 6t$.

39. Taking the Laplace transform of the given differential equation we obtain

$$\mathscr{L}\{y\} = \frac{2w_0}{EIL}\left(\frac{L}{48}\cdot\frac{4!}{s^5} - \frac{1}{120}\cdot\frac{5!}{s^6} + \frac{1}{120}\cdot\frac{5!}{s^6}e^{-sL/2}\right) + \frac{c_1}{2}\cdot\frac{2!}{s^3} + \frac{c_2}{6}\cdot\frac{3!}{s^4}$$

so that

$$y = \frac{2w_0}{EIL}\left[\frac{L}{48}x^4 - \frac{1}{120}x^5 + \frac{1}{120}\left(x-\frac{L}{2}\right)^5\mathcal{U}\left(x-\frac{L}{2}\right) + \frac{c_1}{2}x^2 + \frac{c_2}{6}x^3\right]$$

where $y''(0) = c_1$ and $y'''(0) = c_2$. Using $y''(L) = 0$ and $y'''(L) = 0$ we find

$$c_1 = w_0 L^2/24EI, \qquad c_2 = -w_0 L/4EI.$$

Hence

$$y = \frac{w_0}{12EIL}\left[-\frac{1}{5}x^5 + \frac{L}{2}x^4 - \frac{L^2}{2}x^3 + \frac{L^3}{4}x^2 + \frac{1}{5}\left(x-\frac{L}{2}\right)^5\mathcal{U}\left(x-\frac{L}{2}\right)\right].$$

8 Systems of Linear Differential Equations

Exercises 8.1

3. From $Dx = -y + t$ and $Dy = x - t$ we obtain $y = t - Dx$, $Dy = 1 - D^2x$, and $(D^2 + 1)x = 1 + t$. Then

$$x = c_1 \cos t + c_2 \sin t + 1 + t$$

and

$$y = c_1 \sin t - c_2 \cos t + t - 1.$$

6. From $(D + 1)x + (D - 1)y = 2$ and $3x + (D + 2)y = -1$ we obtain $x = -\frac{1}{3} - \frac{1}{3}(D + 2)y$, $Dx = -\frac{1}{3}(D^2 + 2D)y$, and $(D^2 + 5)y = -7$. Then

$$y = c_1 \cos \sqrt{5}\,t + c_2 \sin \sqrt{5}\,t - \frac{7}{5}$$

and

$$x = \left(-\frac{2}{3}c_1 - \frac{\sqrt{5}}{3}c_2\right) \cos \sqrt{5}\,t + \left(\frac{\sqrt{5}}{3}c_1 - \frac{2}{3}c_2\right) \sin \sqrt{5}\,t + \frac{3}{5}.$$

9. From $Dx + D^2y = e^{3t}$ and $(D + 1)x + (D - 1)y = 4e^{3t}$ we obtain $D(D^2 + 1)x = 34e^{3t}$ and $D(D^2 + 1)y = -8e^{3t}$. Then

$$y = c_1 + c_2 \sin t + c_3 \cos t - \frac{4}{15}e^{3t}$$

and

$$x = c_4 + c_5 \sin t + c_6 \cos t + \frac{17}{15}e^{3t}.$$

Substituting into $(D + 1)x + (D - 1)y = 4e^{3t}$ gives

$$(c_4 - c_1) + (c_5 - c_6 - c_3 - c_2) \sin t + (c_6 + c_5 + c_2 - c_3) \cos t = 0$$

so that $c_4 = c_1$, $c_5 = c_3$, $c_6 = -c_2$, and

$$x = c_1 - c_2 \cos t + c_3 \sin t + \frac{17}{15}e^{3t}.$$

12. From $(2D^2 - D - 1)x - (2D + 1)y = 1$ and $(D - 1)x + Dy = -1$ we obtain $(2D + 1)(D - 1)(D + 1)x = -1$ and $(2D + 1)(D + 1)y = -2$. Then

$$x = c_1 e^{-t/2} + c_2 e^{-t} + c_3 e^{t} + 1$$

and

100

$$y = c_4 e^{-t/2} + c_5 e^{-t} - 2.$$

Substituting into $(D-1)x + Dy = -1$ gives

$$\left(-\frac{3}{2}c_1 - \frac{1}{2}c_4\right)e^{-t/2} + (-2c_2 - c_5)e^{-t} = 0$$

so that $c_4 = -3c_1$, $c_5 = -2c_2$, and

$$y = -3c_1 e^{-t/2} - 2c_2 e^{-t} - 2.$$

15. From $(D-1)x + (D^2+1)y = 1$ and $(D^2-1)x + (D+1)y = 2$ we obtain $D^2(D-1)(D+1)x = 1$ and $D^2(D-1)(D+1)y = 1$. Then

$$x = c_1 + c_2 t + c_3 e^t + c_4 e^{-t} - \frac{1}{2}t^2$$

and

$$y = c_5 + c_6 t + c_7 e^t + c_8 e^{-t} - \frac{1}{2}t^2.$$

Substituting into $(D-1)x + (D^2+1)y = 1$ gives

$$(c_2 - c_1 - 1 + c_5) + (c_6 - c_2 - 1)t + (2c_8 - 2c_4)e^{-t} + (2c_7)e^t = 1$$

so that $c_6 = c_2 + 1$, $c_8 = c_4$, $c_7 = 0$, $c_5 = c_1 - c_2 + 2$, and

$$y = (c_1 - c_2 + 2) + (c_2 + 1)t + c_4 e^{-t} - \frac{1}{2}t^2.$$

18. From $Dx + z = e^t$, $(D-1)x + Dy + Dz = 0$, and $x + 2y + Dz = e^t$ we obtain $z = -Dx + e^t$, $Dz = -D^2 x + e^t$, and the system $(-D^2 + D - 1)x + Dy = -e^t$ and $(-D^2 + 1)x + 2y = 0$. Then $y = \frac{1}{2}(D^2 - 1)x$, $Dy = \frac{1}{2}D(D^2 - 1)x$, and $(D-2)(D^2+1)x = -2e^t$ so that

$$x = c_1 e^{2t} + c_2 \cos t + c_3 \sin t + e^t,$$

$$y = \frac{3}{2}c_1 e^{2t} - c_2 \cos t - c_3 \sin t,$$

and

$$z = -2c_1 e^{2t} - c_3 \cos t + c_2 \sin t.$$

21. From $2Dx + (D-1)y = t$ and $Dx + Dy = t^2$ we obtain $(D+1)y = 2t^2 - t$. Then

$$y = c_1 e^{-t} + 2t^2 - 5t + 5$$

and $Dx = c_1 e^{-t} + t^2 - 4t + 5$ so that

$$x = -c_1 e^{-t} + c_2 + \frac{1}{3}t^3 - 2t^2 + 5t.$$

24. From $Dx - y = -1$ and $3x + (D-2)y = 0$ we obtain $x = -\frac{1}{3}(D-2)y$ so that $Dx = -\frac{1}{3}(D^2 - 2D)y$.
Then $-\frac{1}{3}(D^2 - 2D)y = y - 1$ and $(D^2 - 2D + 3)y = 3$. Thus

$$y = e^t \left(c_1 \cos \sqrt{2}\,t + c_2 \sin \sqrt{2}\,t \right) + 1$$

and

$$x = \frac{1}{3} e^t \left[\left(c_1 - \sqrt{2}\,c_2 \right) \cos \sqrt{2}\,t + \left(\sqrt{2}\,c_1 + c_2 \right) \sin \sqrt{2}\,t \right] + \frac{2}{3}.$$

Using $x(0) = y(0) = 0$ we obtain

$$c_1 + 1 = 0$$

$$\frac{1}{3} \left(c_1 - \sqrt{2}\,c_2 \right) + \frac{2}{3} = 0.$$

Thus $c_1 = -1$ and $c_2 = \sqrt{2}/2$. The solution of the initial value problem is

$$x = e^t \left(-\frac{2}{3} \cos \sqrt{2}\,t - \frac{\sqrt{2}}{6} \sin \sqrt{2}\,t \right) + \frac{2}{3}$$

$$y = e^t \left(-\cos \sqrt{2}\,t + \frac{\sqrt{2}}{2} \sin \sqrt{2}\,t \right) + 1.$$

Exercises 8.2

3. Taking the Laplace transform of the system gives

$$s\mathcal{L}\{x\} + 1 = \mathcal{L}\{x\} - 2\mathcal{L}\{y\}$$

$$s\mathcal{L}\{y\} - 2 = 5\mathcal{L}\{x\} - \mathcal{L}\{y\}$$

so that

$$\mathcal{L}\{x\} = \frac{-s-5}{s^2+9} = -\frac{s}{s^2+9} - \frac{5}{3}\frac{3}{s^2+9}$$

and

$$x = -\cos 3t - \frac{5}{3} \sin 3t.$$

Then

$$y = \frac{1}{2}x - \frac{1}{2}x' = 2\cos 3t - \frac{7}{3} \sin 3t.$$

6. Taking the Laplace transform of the system gives

$$(s+1)\mathcal{L}\{x\} - (s-1)\mathcal{L}\{y\} = -1$$

$$s\mathcal{L}\{x\} + (s+2)\mathcal{L}\{y\} = 1$$

so that

$$\mathscr{L}\{y\} = \frac{s+1/2}{s^2+s+1} = \frac{s+1/2}{(s+1/2)^2+(\sqrt{3}/2)^2}$$

and

$$\mathscr{L}\{x\} = \frac{-3/2}{s^2+s+1} = \frac{-3/2}{(s+1/2)^2+(\sqrt{3}/2)^2}.$$

Then

$$y = e^{-t/2}\cos\frac{\sqrt{3}}{2}t \quad \text{and} \quad x = e^{-t/2}\sin\frac{\sqrt{3}}{2}t.$$

9. Adding the equations and then subtracting them gives

$$\frac{d^2x}{dt^2} = \frac{1}{2}t^2 + 2t$$

$$\frac{d^2y}{dt^2} = \frac{1}{2}t^2 - 2t.$$

Taking the Laplace transform of the system gives

$$\mathscr{L}\{x\} = 8\frac{1}{s} + \frac{1}{24}\frac{4!}{s^5} + \frac{1}{3}\frac{3!}{s^4}$$

and

$$\mathscr{L}\{y\} = \frac{1}{24}\frac{4!}{s^5} - \frac{1}{3}\frac{3!}{s^4}$$

so that

$$x = 8 + \frac{1}{24}t^4 + \frac{1}{3}t^3 \quad \text{and} \quad y = \frac{1}{24}t^4 - \frac{1}{3}t^3.$$

12. Taking the Laplace transform of the system gives

$$(s-4)\,\mathscr{L}\{x\} + 2\mathscr{L}\{y\} = \frac{2e^{-s}}{s}$$

$$-3\,\mathscr{L}\{x\} + (s+1)\,\mathscr{L}\{y\} = \frac{1}{2} + \frac{e^{-s}}{s}$$

so that

$$\mathscr{L}\{x\} = \frac{-1/2}{(s-1)(s-2)} + e^{-s}\frac{1}{(s-1)(s-2)}$$

$$= \left[\frac{1}{2}\frac{1}{s-1} - \frac{1}{2}\frac{1}{s-2}\right] + e^{-s}\left[-\frac{1}{s-1} + \frac{1}{s-2}\right]$$

and

$$\mathscr{L}\{y\} = \frac{e^{-s}}{s} + \frac{s/4-1}{(s-1)(s-2)} + e^{-s}\frac{-s/2+2}{(s-1)(s-2)}$$

$$= \frac{3}{4}\frac{1}{s-1} - \frac{1}{2}\frac{1}{s-2} + e^{-s}\left[\frac{1}{s} - \frac{3}{2}\frac{1}{s-1} + \frac{1}{s-2}\right].$$

Then

$$x = \frac{1}{2}e^t - \frac{1}{2}e^{2t} + \left[-e^{t-1} + e^{2(t-1)}\right]\mathcal{U}(t-1)$$

and

$$y = \frac{3}{4}e^t - \frac{1}{2}e^{2t} + \left[1 - \frac{3}{2}e^{t-1} + e^{2(t-1)}\right]\mathcal{U}(t-1).$$

15. (a) By Kirchoff's first law we have $i_1 = i_2 + i_3$. By Kirchoff's second law, on each loop we have
$E(t) = Ri_1 + L_1 i_2'$ and $E(t) = Ri_1 + L_2 i_3'$ or $L_1 i_2' + Ri_2 + Ri_3 = E(t)$ and $L_2 i_3' + Ri_2 + Ri_3 = E(t)$.

(b) Taking the Laplace transform of the system

$$0.01 i_2' + 5i_2 + 5i_3 = 100$$

$$0.0125 i_3' + 5i_2 + 5i_3 = 100$$

gives

$$(s + 500)\,\mathcal{L}\{i_2\} + 500\mathcal{L}\{i_3\} = \frac{10,000}{s}$$

$$400\mathcal{L}\{i_2\} + (s + 400)\,\mathcal{L}\{i_3\} = \frac{8,000}{s}$$

so that

$$\mathcal{L}\{i_3\} = \frac{8,000}{s^2 + 900s} = \frac{80}{9}\frac{1}{s} - \frac{80}{9}\frac{1}{s+900}.$$

Then

$$i_3 = \frac{80}{9} - \frac{80}{9}e^{-900t} \quad \text{and} \quad i_2 = 20 - 0.0025 i_3' - i_3 = \frac{100}{9} - \frac{100}{9}e^{-900t}.$$

(c) $i_1 = i_2 + i_3 = 20 - 20e^{-900t}$.

18. By Kirchoff's first law we have $i_1 = i_2 + i_3$. By Kirchoff's second law, on each loop we have
$E(t) = Li_1' + Ri_2$ and $E(t) = Li_1' + \frac{1}{C}q$ so that $q = CRi_2$. Then $i_3 = q' = CRi_2'$ so that system is

$$Li' + Ri_2 = E(t)$$

$$CRi_2' + i_2 - i_1 = 0.$$

21. (a) Using Kirchoff's first law we write $i_1 = i_2 + i_3$. Since $i_2 = dq/dt$ we have $i_1 - i_3 = dq/dt$. Using Kirchoff's second law and summing the voltage drops across the shorter loop gives

$$E(t) = iR_1 + \frac{1}{C}q, \tag{1}$$

so that

$$i_1 = \frac{1}{R_1}E(t) - \frac{1}{R_1 C}q.$$

Then

$$\frac{dq}{dt} = i_1 - i_3 = \frac{1}{R_1}E(t) - \frac{1}{R_1 C}q - i_3$$

and

$$R_1\frac{dq}{dt} + \frac{1}{C}q + R_1 i_3 = E(t).$$

Summing the voltage drops across the longer loop gives

$$E(t) = i_1 R_1 + L\frac{di_3}{dt} + R_2 i_3.$$

Combining this with (1) we obtain

$$i_1 R_1 + L\frac{di_3}{dt} + R_2 i_3 = i_1 R_1 + \frac{1}{C}q$$

or

$$L\frac{di_3}{dt} + R_2 i_3 - \frac{1}{C}q = 0.$$

(b) Using $L = R_1 = R_2 = C = 1$, $E(t) = 50e^{-t}\,\mathscr{U}(t-1) = 50e^{-1}e^{-(t-1)}\,\mathscr{U}(t-1)$, $q(0) = i_3(0) = 0$, and taking the Laplace transform of the system we obtain

$$(s+1)\,\mathscr{L}\{q\} + \mathscr{L}\{i_3\} = \frac{50e^{-1}}{s+1}e^{-s}$$

$$(s+1)\,\mathscr{L}\{i_3\} - \mathscr{L}\{q\} = 0,$$

so that

$$\mathscr{L}\{q\} = \frac{50e^{-1}e^{-s}}{(s+1)^2 + 1}$$

and

$$q(t) = 50e^{-1}e^{-(t-1)}\sin(t-1)\mathscr{U}(t-1) = 50e^{-t}\sin(t-1)\mathscr{U}(t-1).$$

Exercises 8.3

3. Let $x_1 = y$, $x_2 = y'$, $x_3 = y''$, and $y''' = 3y'' - 6y' + 10y + t^2 + 1$ so that

$$x_1' = x_2$$

$$x_2' = x_3$$

$$x_3' = 3x_3 - 6x_2 + 10x_1 + t^2 + 1.$$

6. Let $x_1 = y$, $x_2 = y'$, $x_3 = y''$, $x_4 = y'''$, and $y^{(4)} = -\frac{1}{2}y''' + 4y + 10$ so that

$$x_1' = x_2$$

$$x_2' = x_3$$

$$x_3' = x_4$$

$$x_4' = -\frac{1}{2}x_4 + 4x_1 + 10.$$

9. From

$$x' + 4x - y' = 7t$$

$$x' + y' - 2y = 3t$$

we obtain

$$2x' + 4x - 2y = 10t$$

$$2y' - 4x - 2y = -4t$$

so that

$$x' = -2x + y + 5t$$

$$y' = 2x + y - 2t.$$

12. From $x'' - 2y'' = \sin t$ and $x'' + y'' = \cos t$ we obtain

$$3x'' = 2\cos t + \sin t$$

$$3y'' = \cos t - \sin t.$$

Let $x_1 = x$, $x_2 = x'$, $x_3 = y$, and $x_4 = y'$. Then

$$x_1' = x_2$$

$$x_2' = \frac{2}{3}\cos t + \frac{1}{3}\sin t$$

$$x_3' = x_4$$

$$x_4' = \frac{1}{3}\cos t - \frac{1}{3}\sin t.$$

15. Let $z_1 = x$, $z_2 = x'$, $z_3 = x''$, $z_4 = y$, and $z_5 = y'$ so that

$$z_1' = z_2$$

$$z_2' = z_3$$

$$z_3' = 4z_1 - 3z_3 + 4z_5$$

$$z_4' = z_5$$

$$z_5' = 10t^2 - 4z_2 + 3z_5.$$

18. The system is

$$x_1' = 2 \cdot 3 + \frac{1}{50}x_2 - \frac{1}{50}x_1 \cdot 4$$

$$x_2' = \frac{1}{50}x_1 \cdot 4 - \frac{1}{50}x_2 - \frac{1}{50}x_2 \cdot 3.$$

21. Since $Dx + Dy = -x - y$ and $Dx + Dy = -\frac{1}{2}y$ we obtain $y = -2x$ and $Dx = -x$. Then $x = c_1 e^{-t}$ and $y = -2c_1 e^{-t}$.

Exercises 8.4

3. (a) $\mathbf{AB} = \begin{pmatrix} -2-9 & 12-6 \\ 5+12 & -30+8 \end{pmatrix} = \begin{pmatrix} -11 & 6 \\ 17 & -22 \end{pmatrix}$

(b) $\mathbf{BA} = \begin{pmatrix} -2-30 & 3+24 \\ 6-10 & -9+8 \end{pmatrix} = \begin{pmatrix} -32 & 27 \\ -4 & -1 \end{pmatrix}$

(c) $\mathbf{A}^2 = \begin{pmatrix} 4+15 & -6-12 \\ -10-20 & 15+16 \end{pmatrix} = \begin{pmatrix} 19 & -18 \\ -30 & 31 \end{pmatrix}$

(d) $\mathbf{B}^2 = \begin{pmatrix} 1+18 & -6+12 \\ -3+6 & 18+4 \end{pmatrix} = \begin{pmatrix} 19 & 6 \\ 3 & 22 \end{pmatrix}$

6. (a) $\mathbf{AB} = \begin{pmatrix} 5 & -6 & 7 \end{pmatrix} \begin{pmatrix} 3 \\ 4 \\ -1 \end{pmatrix} = (-16)$

(b) $\mathbf{BA} = \begin{pmatrix} 3 \\ 4 \\ -1 \end{pmatrix} \begin{pmatrix} 5 & -6 & 7 \end{pmatrix} = \begin{pmatrix} 15 & -18 & 21 \\ 20 & -24 & 28 \\ -5 & 6 & -7 \end{pmatrix}$

(c) $(\mathbf{BA})\mathbf{C} = \begin{pmatrix} 15 & -18 & 21 \\ 20 & -24 & 28 \\ -5 & 6 & -7 \end{pmatrix} \begin{pmatrix} 1 & 2 & 4 \\ 0 & 1 & -1 \\ 3 & 2 & 1 \end{pmatrix} = \begin{pmatrix} 78 & 54 & 99 \\ 104 & 72 & 132 \\ -26 & -18 & -33 \end{pmatrix}$

(d) Since $\mathbf{AB}$ is 1×1 and $\mathbf{C}$ is 3×3 the product $(\mathbf{AB})\mathbf{C}$ is not defined.

9. (a) $(\mathbf{AB})^T = \begin{pmatrix} 7 & 10 \\ 38 & 75 \end{pmatrix}^T = \begin{pmatrix} 7 & 38 \\ 10 & 75 \end{pmatrix}$

(b) $\mathbf{B}^T\mathbf{A}^T = \begin{pmatrix} 5 & -2 \\ 10 & -5 \end{pmatrix} \begin{pmatrix} 3 & 8 \\ 4 & 1 \end{pmatrix} = \begin{pmatrix} 7 & 38 \\ 10 & 75 \end{pmatrix}$

12. $\begin{pmatrix} 6t \\ 3t^2 \\ -3t \end{pmatrix} + \begin{pmatrix} -t+1 \\ -t^2+t \\ 3t-3 \end{pmatrix} - \begin{pmatrix} 6t \\ 8 \\ -10t \end{pmatrix} = \begin{pmatrix} -t+1 \\ 2t^2+t-8 \\ 10t-3 \end{pmatrix}$

15. Since $\det \mathbf{A} = 0$, $\mathbf{A}$ is singular.

18. Since $\det \mathbf{A} = -6$, $\mathbf{A}$ is nonsingular.

$$\mathbf{A}^{-1} = -\frac{1}{6}\begin{pmatrix} 2 & -10 \\ -2 & 7 \end{pmatrix}$$

21. Since $\det \mathbf{A} = -9$, $\mathbf{A}$ is nonsingular. The cofactors are

$$\begin{array}{lll} A_{11} = -2 & A_{12} = -13 & A_{13} = 8 \\ A_{21} = -2 & A_{22} = 5 & A_{23} = -1 \\ A_{31} = -1 & A_{32} = 7 & A_{33} = -5. \end{array}$$

Then

$$\mathbf{A}^{-1} = -\frac{1}{9}\begin{pmatrix} -2 & -13 & 8 \\ -2 & 5 & -1 \\ -1 & 7 & -5 \end{pmatrix}^T = -\frac{1}{9}\begin{pmatrix} -2 & -2 & -1 \\ -13 & 5 & 7 \\ 8 & -1 & -5 \end{pmatrix}$$

24. Since $\det \mathbf{A}(t) = 2e^{2t} \neq 0$, $\mathbf{A}$ is nonsingular.

$$\mathbf{A}^{-1} = \frac{1}{2}e^{-2t}\begin{pmatrix} e^t \sin t & 2e^t \cos t \\ -e^t \cos t & 2e^t \sin t \end{pmatrix}$$

27. $\mathbf{X} = \begin{pmatrix} 2e^{2t} + 8e^{-3t} \\ -2e^{2t} + 4e^{-3t} \end{pmatrix}$ so that $\dfrac{d\mathbf{X}}{dt} = \begin{pmatrix} 4e^{2t} - 24e^{-3t} \\ -4e^{2t} - 12e^{-3t} \end{pmatrix}$.

30. (a) $\dfrac{d\mathbf{A}}{dt} = \begin{pmatrix} -2t/(t^2+1)^2 & 3 \\ 2t & 1 \end{pmatrix}$

(b) $\dfrac{d\mathbf{B}}{dt} = \begin{pmatrix} 6 & 0 \\ -1/t^2 & 4 \end{pmatrix}$

(c) $\displaystyle\int_0^1 \mathbf{A}(t)\,dt = \begin{pmatrix} \tan^{-1} t & \frac{3}{2}t^2 \\ \frac{1}{3}t^3 & \frac{1}{2}t^2 \end{pmatrix} \bigg|_{t=0}^{t=1} = \begin{pmatrix} \frac{\pi}{4} & \frac{3}{2} \\ \frac{1}{3} & \frac{1}{2} \end{pmatrix}$

(d) $\displaystyle\int_1^2 \mathbf{B}(t)\,dt = \begin{pmatrix} 3t^2 & 2t \\ \ln t & 2t^2 \end{pmatrix} \bigg|_{t=1}^{t=2} = \begin{pmatrix} 9 & 2 \\ \ln 2 & 6 \end{pmatrix}$

(e) $\mathbf{A}(t)\mathbf{B}(t) = \begin{pmatrix} 6t/(t^2+1)+3 & 2/(t^2+1)+12t^2 \\ 6t^3+1 & 2t^2+4t^2 \end{pmatrix}$

(f) $\dfrac{d}{dt}\mathbf{A}(t)\mathbf{B}(t) = \begin{pmatrix} (6-6t^2)/(t^2+1)^2 & -4t/(t^2+1)^2+24t \\ 18t^2 & 12t \end{pmatrix}$

(g) $\displaystyle\int_1^t \mathbf{A}(s)\mathbf{B}(s)\,ds = \begin{pmatrix} 6s/(s^2+1)+3 & 2/(s^2+1)+12s^2 \\ 6s^3+1 & 6s^2 \end{pmatrix} \bigg|_{s=1}^{s=t}$

$= \begin{pmatrix} 3t+3\ln(t^2+1)-3-3\ln 2 & 4t^3+2\tan^{-1}t-4-\pi/2 \\ (3/2)t^4+t-(5/2) & 2t^3-2 \end{pmatrix}$

33. $\begin{pmatrix} 1 & -1 & -5 & | & 7 \\ 5 & 4 & -16 & | & -10 \\ 0 & 1 & 1 & | & -5 \end{pmatrix} \Longrightarrow \begin{pmatrix} 1 & -1 & -5 & | & 7 \\ 0 & 1 & 1 & | & -5 \\ 0 & 9 & 9 & | & -45 \end{pmatrix} \Longrightarrow \begin{pmatrix} 1 & 0 & -4 & | & 2 \\ 0 & 1 & 1 & | & -5 \\ 0 & 0 & 0 & | & 0 \end{pmatrix}$

Letting $z = t$ we find $y = -5 - t$, and $x = 2 + 4t$.

36. $\begin{pmatrix} 1 & 0 & 2 & | & 8 \\ 1 & 2 & -2 & | & 4 \\ 2 & 5 & -6 & | & 6 \end{pmatrix} \Longrightarrow \begin{pmatrix} 1 & 0 & 2 & | & 8 \\ 0 & 2 & -4 & | & -4 \\ 0 & 5 & -10 & | & -10 \end{pmatrix} \Longrightarrow \begin{pmatrix} 1 & 0 & 2 & | & 8 \\ 0 & 1 & -2 & | & -2 \\ 0 & 0 & 0 & | & 0 \end{pmatrix}$

Letting $z = t$ we find $y = -2 + 2t$, and $x = 8 - 2t$.

39. $\begin{pmatrix} 1 & 2 & 4 & | & 2 \\ 2 & 4 & 3 & | & 1 \\ 1 & 2 & -1 & | & 7 \end{pmatrix} \Longrightarrow \begin{pmatrix} 1 & 2 & 4 & | & 2 \\ 0 & 0 & -5 & | & -3 \\ 0 & 0 & -5 & | & 5 \end{pmatrix} \Longrightarrow \begin{pmatrix} 1 & 2 & 0 & | & -2/5 \\ 0 & 0 & 1 & | & 3/5 \\ 0 & 0 & 0 & | & 8 \end{pmatrix}$

There is no solution.

42. We solve

$$\det(\mathbf{A} - \lambda\mathbf{I}) = \begin{vmatrix} 2-\lambda & 1 \\ 2 & 1-\lambda \end{vmatrix} = \lambda(\lambda-3) = 0.$$

For $\lambda_1 = 0$ we have

$$\begin{pmatrix} 2 & 1 & | & 0 \\ 2 & 1 & | & 0 \end{pmatrix} \implies \begin{pmatrix} 1 & 1/2 & | & 0 \\ 0 & 0 & | & 0 \end{pmatrix}$$

so that $k_1 = -\frac{1}{2}k_2$. If $k_2 = 2$ then

$$\mathbf{K}_1 = \begin{pmatrix} -1 \\ 2 \end{pmatrix}.$$

For $\lambda_2 = 3$ we have

$$\begin{pmatrix} -1 & 1 & | & 0 \\ 2 & -2 & | & 0 \end{pmatrix} \implies \begin{pmatrix} 1 & -1 & | & 0 \\ 0 & 0 & | & 0 \end{pmatrix}$$

so that $k_1 = k_2$. If $k_2 = 1$ then

$$\mathbf{K}_2 = \begin{pmatrix} 1 \\ 1 \end{pmatrix}.$$

45. We solve

$$\det(\mathbf{A} - \lambda\mathbf{I}) = \begin{vmatrix} 5-\lambda & -1 & 0 \\ 0 & -5-\lambda & 9 \\ 5 & -1 & -\lambda \end{vmatrix} = \begin{vmatrix} 4-\lambda & -1 & 0 \\ 4-\lambda & -5-\lambda & 9 \\ 4-\lambda & -1 & -\lambda \end{vmatrix} = \lambda(4-\lambda)(\lambda+4) = 0.$$

If $\lambda_1 = 0$ then

$$\begin{pmatrix} 5 & -1 & 0 & | & 0 \\ 0 & -5 & 9 & | & 0 \\ 5 & -1 & 0 & | & 0 \end{pmatrix} \implies \begin{pmatrix} 1 & 0 & -9/25 & | & 0 \\ 0 & 1 & -9/5 & | & 0 \\ 0 & 0 & 0 & | & 0 \end{pmatrix}$$

so that $k_1 = \frac{9}{25}k_3$ and $k_2 = \frac{9}{5}k_3$. If $k_3 = 25$ then

$$\mathbf{K}_1 = \begin{pmatrix} 9 \\ 45 \\ 25 \end{pmatrix}.$$

If $\lambda_2 = 4$ then

$$\begin{pmatrix} 1 & -1 & 0 & | & 0 \\ 0 & -9 & 9 & | & 0 \\ 5 & -1 & -4 & | & 0 \end{pmatrix} \implies \begin{pmatrix} 1 & 0 & -1 & | & 0 \\ 0 & 1 & -1 & | & 0 \\ 0 & 0 & 0 & | & 0 \end{pmatrix}$$

so that $k_1 = k_3$ and $k_2 = k_3$. If $k_3 = 1$ then

$$\mathbf{K}_2 = \begin{pmatrix} 1 \\ 1 \\ 1 \end{pmatrix}.$$

If $\lambda_3 = -4$ then

$$\begin{pmatrix} 9 & -1 & 0 & | & 0 \\ 0 & -1 & 9 & | & 0 \\ 5 & -1 & 4 & | & 0 \end{pmatrix} \implies \begin{pmatrix} 1 & 0 & -1 & | & 0 \\ 0 & 1 & -9 & | & 0 \\ 0 & 0 & 0 & | & 0 \end{pmatrix}$$

so that $k_1 = k_3$ and $k_2 = 9k_3$. If $k_3 = 1$ then

$$\mathbf{K}_3 = \begin{pmatrix} 1 \\ 9 \\ 1 \end{pmatrix}.$$

48. We solve

$$\det(\mathbf{A} - \lambda\mathbf{I}) = \begin{vmatrix} 1-\lambda & 6 & 0 \\ 0 & 2-\lambda & 1 \\ 0 & 1 & 2-\lambda \end{vmatrix} = \begin{vmatrix} 1-\lambda & 6 & 0 \\ 0 & 3-\lambda & 3-\lambda \\ 0 & 1 & 2-\lambda \end{vmatrix} = (3-\lambda)(1-\lambda)^2 = 0.$$

For $\lambda = 3$ we have

$$\begin{pmatrix} -2 & 6 & 0 & | & 0 \\ 0 & 0 & 0 & | & 0 \\ 0 & 1 & -1 & | & 0 \end{pmatrix} \implies \begin{pmatrix} 1 & 0 & -3 & | & 0 \\ 0 & 1 & -1 & | & 0 \\ 0 & 0 & 0 & | & 0 \end{pmatrix}$$

so that $k_1 = 3k_3$ and $k_2 = k_3$. If $k_3 = 1$ then

$$\mathbf{K}_1 = \begin{pmatrix} 3 \\ 1 \\ 1 \end{pmatrix}.$$

For $\lambda_2 = \lambda_3 = 1$ we have

$$\begin{pmatrix} 0 & 6 & 0 & | & 0 \\ 0 & 1 & 1 & | & 0 \\ 0 & 1 & 1 & | & 0 \end{pmatrix} \implies \begin{pmatrix} 0 & 1 & 0 & | & 0 \\ 0 & 0 & 1 & | & 0 \\ 0 & 0 & 0 & | & 0 \end{pmatrix}$$

so that $k_2 = 0$ and $k_3 = 0$. If $k_1 = 1$ then

$$\mathbf{K}_2 = \begin{pmatrix} 1 \\ 0 \\ 0 \end{pmatrix}.$$

51. Let

$$\mathbf{A} = \begin{pmatrix} a_{11} & a_{12} \\ a_{21} & a_{22} \end{pmatrix}.$$

Then

$$\frac{d}{dt}[\mathbf{A}(t)\mathbf{X}(t)] = \frac{d}{dt}\begin{pmatrix} a_1 & a_2 \\ a_3 & a_4 \end{pmatrix}\begin{pmatrix} x_1 \\ x_2 \end{pmatrix} = \frac{d}{dt}\begin{pmatrix} a_1 x_1 + a_2 x_2 \\ a_3 x_1 + a_4 x_2 \end{pmatrix} = \begin{pmatrix} a_1 x_1' + a_1' x_1 + a_2 x_2' + a_2' x_2 \\ a_3 x_1' + a_3' x_1 + a_4 x_2' + a_4' x_2 \end{pmatrix}$$

$$= \begin{pmatrix} a_1 & a_2 \\ a_3 & a_4 \end{pmatrix}\begin{pmatrix} x_1' \\ x_2' \end{pmatrix} + \begin{pmatrix} a_1' & a_2' \\ a_3' & a_4' \end{pmatrix}\begin{pmatrix} x_1 \\ x_2 \end{pmatrix} = \mathbf{A}(t)\mathbf{X}'(t) + \mathbf{A}'(t)\mathbf{X}(t).$$

54. Since

$$(\mathbf{AB})(\mathbf{B}^{-1}\mathbf{A}^{-1}) = \mathbf{A}(\mathbf{BB}^{-1})\mathbf{A}^{-1} = \mathbf{AIA}^{-1} = \mathbf{AA}^{-1} = \mathbf{I}$$

and

$$(\mathbf{B}^{-1}\mathbf{A}^{-1})(\mathbf{AB}) = \mathbf{B}^{-1}(\mathbf{A}^{-1}\mathbf{A})\mathbf{B} = \mathbf{B}^{-1}\mathbf{IB} = \mathbf{B}^{-1}\mathbf{B} = \mathbf{I}$$

we have

$$(\mathbf{AB})^{-1} = \mathbf{B}^{-1}\mathbf{A}^{-1}.$$

Exercises 8.5

3. Let $\mathbf{X} = \begin{pmatrix} x \\ y \\ z \end{pmatrix}$. Then

$$\mathbf{X}' = \begin{pmatrix} -3 & 4 & -9 \\ 6 & -1 & 0 \\ 10 & 4 & 3 \end{pmatrix}\mathbf{X}.$$

6. Let $\mathbf{X} = \begin{pmatrix} x \\ y \end{pmatrix}$. Then

$$\mathbf{X}' = \begin{pmatrix} -3 & 4 \\ 5 & 9 \end{pmatrix}\mathbf{X} + \begin{pmatrix} e^{-t}\sin 2t \\ 4e^{-t}\cos 2t \end{pmatrix}.$$

9. $\dfrac{dx}{dt} = x - y + 2z + e^{-t} - 3t;\quad \dfrac{dy}{dt} = 3x - 4y + z + 2e^{-t} + t;\quad \dfrac{dz}{dt} = -2x + 5y + 6z + 2e^{-t} - t$

12. Since

$$\mathbf{X}' = \begin{pmatrix} 5\cos t - 5\sin t \\ 2\cos t - 4\sin t \end{pmatrix}e^t \quad \text{and} \quad \begin{pmatrix} -2 & 5 \\ -2 & 4 \end{pmatrix}\mathbf{X} = \begin{pmatrix} 5\cos t - 5\sin t \\ 2\cos t - 4\sin t \end{pmatrix}e^t$$

we see that

$$\mathbf{X}' = \begin{pmatrix} -2 & 5 \\ -2 & 4 \end{pmatrix}\mathbf{X}.$$

15. Since

$$\mathbf{X}' = \begin{pmatrix} 0 \\ 0 \\ 0 \end{pmatrix} \quad \text{and} \quad \begin{pmatrix} 1 & 2 & 1 \\ 6 & -1 & 0 \\ -1 & -2 & -1 \end{pmatrix} \mathbf{X} = \begin{pmatrix} 0 \\ 0 \\ 0 \end{pmatrix}$$

we see that

$$\mathbf{X}' = \begin{pmatrix} 1 & 2 & 1 \\ 6 & -1 & 0 \\ -1 & -2 & -1 \end{pmatrix} \mathbf{X}.$$

18. Yes, since $W(\mathbf{X}_1, \mathbf{X}_2) = 8e^{2t} \neq 0$ and $\mathbf{X}_1$ and $\mathbf{X}_2$ are linearly independent on $-\infty < t < \infty$.

21. Since

$$\mathbf{X}_p' = \begin{pmatrix} 2 \\ -1 \end{pmatrix} \quad \text{and} \quad \begin{pmatrix} 1 & 4 \\ 3 & 2 \end{pmatrix} \mathbf{X}_p + \begin{pmatrix} 2 \\ -4 \end{pmatrix} t + \begin{pmatrix} -7 \\ -18 \end{pmatrix} = \begin{pmatrix} 2 \\ -1 \end{pmatrix}$$

we see that

$$\mathbf{X}_p' = \begin{pmatrix} 1 & 4 \\ 3 & 2 \end{pmatrix} \mathbf{X}_p + \begin{pmatrix} 2 \\ -4 \end{pmatrix} t + \begin{pmatrix} -7 \\ -18 \end{pmatrix}.$$

24. Since

$$\mathbf{X}_p' = \begin{pmatrix} 3\cos 3t \\ 0 \\ -3\sin 3t \end{pmatrix} \quad \text{and} \quad \begin{pmatrix} 1 & 2 & 3 \\ -4 & 2 & 0 \\ -6 & 1 & 0 \end{pmatrix} \mathbf{X}_p + \begin{pmatrix} -1 \\ 4 \\ 3 \end{pmatrix} \sin 3t = \begin{pmatrix} 3\cos 3t \\ 0 \\ -3\sin 3t \end{pmatrix}$$

we see that

$$\mathbf{X}_p' = \begin{pmatrix} 1 & 2 & 3 \\ -4 & 2 & 0 \\ -6 & 1 & 0 \end{pmatrix} \mathbf{X}_p + \begin{pmatrix} -1 \\ 4 \\ 3 \end{pmatrix} \sin 3t.$$

27. $\Phi(t) = \begin{pmatrix} e^{2t} & e^{7t} \\ -2e^{2t} & 3e^{7t} \end{pmatrix}$ and $\Phi^{-1}(t) = \dfrac{1}{5e^{9t}} \begin{pmatrix} 3e^{7t} & -e^{7t} \\ 2e^{2t} & e^{2t} \end{pmatrix}.$

30. $\Phi(t) = \begin{pmatrix} 2\cos t & -2\sin t \\ 3\cos t + \sin t & -3\sin t + \cos t \end{pmatrix}$ and $\Phi^{-1}(t) = \dfrac{1}{2} \begin{pmatrix} -3\sin t + \cos t & 2\sin t \\ -3\cos t - \sin t & 2\cos t \end{pmatrix}.$

33. We have

$$\mathbf{X}(t) = c_1 \begin{pmatrix} -1 \\ 3 \end{pmatrix} e^t + c_2 \begin{pmatrix} -1 \\ 3 \end{pmatrix} te^t + c_2 \begin{pmatrix} 0 \\ -1 \end{pmatrix} e^t$$

so that

$$\mathbf{X}(0) = c_1 \begin{pmatrix} -1 \\ 3 \end{pmatrix} + c_2 \begin{pmatrix} -1 \\ 3 \end{pmatrix} + c_2 \begin{pmatrix} 0 \\ -1 \end{pmatrix} = \begin{pmatrix} 1 \\ 0 \end{pmatrix}.$$

and

$$\mathbf{X}(0) = c_1 \begin{pmatrix} -1 \\ 3 \end{pmatrix} + c_2 \begin{pmatrix} -1 \\ 3 \end{pmatrix} + c_2 \begin{pmatrix} 0 \\ -1 \end{pmatrix} = \begin{pmatrix} 0 \\ 1 \end{pmatrix}$$

which give $c_1 = -1$, $c_2 = -3$, and $c_1 = 0$, $c_2 = -1$. Then

$$\Psi(t) = \begin{pmatrix} 3te^t + e^t & te^t \\ -9te^t & -3te^t + e^t \end{pmatrix}.$$

36. Since the column vectors of $\Psi(t)$ solve $\mathbf{X}' = \mathbf{AX}$ we know that $\mathbf{X} = \Psi(t)\mathbf{X}_0$ solves $\mathbf{X}' = \mathbf{AX}$, and $\mathbf{X}(t_0) = \Psi(t_0)\mathbf{X}_0 = \mathbf{IX}_0 = \mathbf{X}_0$.

Exercises 8.6

3. The system is

$$\mathbf{X}' = \begin{pmatrix} -4 & 2 \\ -5/2 & 2 \end{pmatrix} \mathbf{X}$$

and $\det(\mathbf{A} - \lambda\mathbf{I}) = (\lambda - 1)(\lambda + 3) = 0$. For $\lambda_1 = 1$ we obtain

$$\begin{pmatrix} -5 & 2 & | & 0 \\ -5/2 & 1 & | & 0 \end{pmatrix} \implies \begin{pmatrix} -5 & 2 & | & 0 \\ 0 & 0 & | & 0 \end{pmatrix} \quad \text{so that} \quad \mathbf{K}_1 = \begin{pmatrix} 2 \\ 5 \end{pmatrix}.$$

For $\lambda_2 = -3$ we obtain

$$\begin{pmatrix} -1 & 2 & | & 0 \\ -5/2 & 5 & | & 0 \end{pmatrix} \implies \begin{pmatrix} -1 & 2 & | & 0 \\ 0 & 0 & | & 0 \end{pmatrix} \quad \text{so that} \quad \mathbf{K}_2 = \begin{pmatrix} 2 \\ 1 \end{pmatrix}.$$

Then

$$\mathbf{X} = c_1 \begin{pmatrix} 2 \\ 5 \end{pmatrix} e^t + c_2 \begin{pmatrix} 2 \\ 1 \end{pmatrix} e^{-3t}.$$

6. The system is

$$\mathbf{X}' = \begin{pmatrix} -6 & 2 \\ -3 & 1 \end{pmatrix} \mathbf{X}$$

and $\det(\mathbf{A} - \lambda\mathbf{I}) = \lambda(\lambda + 5) = 0$. For $\lambda_1 = 0$ we obtain

$$\begin{pmatrix} -6 & 2 & | & 0 \\ -3 & 1 & | & 0 \end{pmatrix} \implies \begin{pmatrix} 1 & -1/3 & | & 0 \\ 0 & 0 & | & 0 \end{pmatrix} \quad \text{so that} \quad \mathbf{K}_1 = \begin{pmatrix} 1 \\ 3 \end{pmatrix}.$$

For $\lambda_2 = -5$ we obtain

$$\begin{pmatrix} -1 & 2 & | & 0 \\ -3 & 6 & | & 0 \end{pmatrix} \implies \begin{pmatrix} 1 & -2 & | & 0 \\ 0 & 0 & | & 0 \end{pmatrix} \quad \text{so that} \quad \mathbf{K}_2 = \begin{pmatrix} 2 \\ 1 \end{pmatrix}.$$

Then

$$\mathbf{X} = c_1 \begin{pmatrix} 1 \\ 3 \end{pmatrix} + c_2 \begin{pmatrix} 2 \\ 1 \end{pmatrix} e^{-5t}.$$

9. We have $\det(\mathbf{A} - \lambda\mathbf{I}) = -(\lambda + 1)(\lambda - 3)(\lambda + 2) = 0$. For $\lambda_1 = -1$, $\lambda_2 = 3$, and $\lambda_3 = -2$ we obtain

$$\mathbf{K}_1 = \begin{pmatrix} -1 \\ 0 \\ 1 \end{pmatrix}, \quad \mathbf{K}_2 = \begin{pmatrix} 1 \\ 4 \\ 3 \end{pmatrix}, \quad \text{and} \quad \mathbf{K}_3 = \begin{pmatrix} 1 \\ -1 \\ 3 \end{pmatrix},$$

so that

$$\mathbf{X} = c_1 \begin{pmatrix} -1 \\ 0 \\ 1 \end{pmatrix} e^{-t} + c_2 \begin{pmatrix} 1 \\ 4 \\ 3 \end{pmatrix} e^{3t} + c_3 \begin{pmatrix} 1 \\ -1 \\ 3 \end{pmatrix} e^{-2t}.$$

12. We have $\det(\mathbf{A} - \lambda\mathbf{I}) = (\lambda - 3)(\lambda + 5)(6 - \lambda) = 0$. For $\lambda_1 = 3$, $\lambda_2 = -5$, and $\lambda_3 = 6$ we obtain

$$\mathbf{K}_1 = \begin{pmatrix} 1 \\ 1 \\ 0 \end{pmatrix}, \quad \mathbf{K}_2 = \begin{pmatrix} 1 \\ -1 \\ 0 \end{pmatrix}, \quad \text{and} \quad \mathbf{K}_3 = \begin{pmatrix} 2 \\ -2 \\ 11 \end{pmatrix},$$

so that

$$\mathbf{X} = c_1 \begin{pmatrix} 1 \\ 1 \\ 0 \end{pmatrix} e^{3t} + c_2 \begin{pmatrix} 1 \\ -1 \\ 0 \end{pmatrix} e^{-5t} + c_3 \begin{pmatrix} 2 \\ -2 \\ 11 \end{pmatrix} e^{6t}.$$

In Problems 15-27 the form of the answer will vary according to the choice of eigenvector. For example, in Problem 15, if $\mathbf{K}_1$ is chosen to be $\begin{pmatrix} 1 \\ -2i \end{pmatrix}$ the solution has the form

$$\mathbf{X} = c_1 \begin{pmatrix} \cos t \\ 2\cos t + \sin t \end{pmatrix} e^{4t} + c_2 \begin{pmatrix} \sin t \\ 2\sin t - \cos t \end{pmatrix} e^{4t}.$$

15. We have $\det(\mathbf{A} - \lambda\mathbf{I}) = \lambda^2 - 8\lambda + 17 = 0$. For $\lambda_1 = 4 + i$ we obtain

$$\mathbf{K}_1 = \begin{pmatrix} 2 + i \\ 5 \end{pmatrix}$$

so that

$$\mathbf{X}_1 = \begin{pmatrix} 2 + i \\ 5 \end{pmatrix} e^{(4+i)t} = \begin{pmatrix} 2\cos t - \sin t \\ 5\cos t \end{pmatrix} e^{4t} + i \begin{pmatrix} \cos t + 2\sin t \\ 5\sin t \end{pmatrix} e^{4t}.$$

Then

$$\mathbf{X} = c_1 \begin{pmatrix} 2\cos t - \sin t \\ 5\cos t \end{pmatrix} e^{4t} + c_2 \begin{pmatrix} 2\sin t + \cos t \\ 5\sin t \end{pmatrix} e^{4t}.$$

18. We have $\det(\mathbf{A} - \lambda\mathbf{I}) = \lambda^2 - 10\lambda + 34 = 0$. For $\lambda_1 = 5 + 3i$ we obtain

$$\mathbf{K}_1 = \begin{pmatrix} 1 - 3i \\ 2 \end{pmatrix}$$

so that

$$\mathbf{X}_1 = \begin{pmatrix} 1 - 3i \\ 2 \end{pmatrix} e^{(5+3i)t} = \begin{pmatrix} \cos 3t + 3\sin 3t \\ 2\cos 3t \end{pmatrix} e^{5t} + i \begin{pmatrix} \sin 3t - 3\cos 3t \\ 2\cos 3t \end{pmatrix} e^{5t}.$$

Then

$$\mathbf{X} = c_1 \begin{pmatrix} \cos 3t + 3\sin 3t \\ 2\cos 3t \end{pmatrix} e^{5t} + c_2 \begin{pmatrix} \sin 3t - 3\cos 3t \\ 2\cos 3t \end{pmatrix} e^{5t}.$$

21. We have $\det(\mathbf{A} - \lambda\mathbf{I}) = -\lambda\left(\lambda^2 + 1\right) = 0$. For $\lambda_1 = 0$ we obtain

$$\mathbf{K}_1 = \begin{pmatrix} 1 \\ 0 \\ 0 \end{pmatrix}.$$

For $\lambda_2 = i$ we obtain

$$\mathbf{K}_2 = \begin{pmatrix} -i \\ i \\ 1 \end{pmatrix}$$

so that

$$\mathbf{X}_2 = \begin{pmatrix} -i \\ i \\ 1 \end{pmatrix} e^{it} = \begin{pmatrix} \sin t \\ -\sin t \\ \cos t \end{pmatrix} + i \begin{pmatrix} -\cos t \\ \cos t \\ \sin t \end{pmatrix}.$$

Then

$$\mathbf{X} = c_1 \begin{pmatrix} 1 \\ 0 \\ 0 \end{pmatrix} + c_2 \begin{pmatrix} \sin t \\ -\sin t \\ \cos t \end{pmatrix} + c_3 \begin{pmatrix} -\cos t \\ \cos t \\ \sin t \end{pmatrix}.$$

24. We have $\det(\mathbf{A} - \lambda\mathbf{I}) = -(\lambda - 6)(\lambda^2 - 8\lambda + 20) = 0$. For $\lambda_1 = 6$ we obtain

$$\mathbf{K}_1 = \begin{pmatrix} 0 \\ 1 \\ 0 \end{pmatrix}.$$

For $\lambda_2 = 4 + 2i$ we obtain

$$\mathbf{K}_2 = \begin{pmatrix} -i \\ 0 \\ 2 \end{pmatrix}$$

so that

$$\mathbf{X}_2 = \begin{pmatrix} -i \\ 0 \\ 2 \end{pmatrix} e^{(4+2i)t} = \begin{pmatrix} \sin 2t \\ 0 \\ 2\cos 2t \end{pmatrix} e^{4t} + i \begin{pmatrix} -\cos 2t \\ 0 \\ 2\sin 2t \end{pmatrix} e^{4t}.$$

Then

$$\mathbf{X} = c_1 \begin{pmatrix} 0 \\ 1 \\ 0 \end{pmatrix} e^{6t} + c_2 \begin{pmatrix} \sin 2t \\ 0 \\ 2\cos 2t \end{pmatrix} e^{4t} + c_3 \begin{pmatrix} -\cos 2t \\ 0 \\ 2\sin 2t \end{pmatrix} e^{4t}.$$

27. We have $\det(\mathbf{A} - \lambda\mathbf{I}) = (1 - \lambda)(\lambda^2 + 25) = 0$. For $\lambda_1 = 1$ we obtain

$$\mathbf{K}_1 = \begin{pmatrix} 25 \\ -7 \\ 6 \end{pmatrix}.$$

For $\lambda_2 = 5i$ we obtain

$$\mathbf{K}_2 = \begin{pmatrix} 1 + 5i \\ 1 \\ 1 \end{pmatrix}$$

so that

$$\mathbf{X}_2 = \begin{pmatrix} 1 + 5i \\ 1 \\ 1 \end{pmatrix} e^{5it} = \begin{pmatrix} \cos 5t - 5\sin 5t \\ \cos 5t \\ \cos 5t \end{pmatrix} + i \begin{pmatrix} \sin 5t + 5\cos 5t \\ \sin 5t \\ \sin 5t \end{pmatrix}.$$

Then

$$\mathbf{X} = c_1 \begin{pmatrix} 25 \\ -7 \\ 6 \end{pmatrix} e^t + c_2 \begin{pmatrix} \cos 5t - 5\sin 5t \\ \cos 5t \\ \cos 5t \end{pmatrix} + c_3 \begin{pmatrix} \sin 5t + 5\cos 5t \\ \sin 5t \\ \sin 5t \end{pmatrix}.$$

If

$$\mathbf{X}(0) = \begin{pmatrix} 4 \\ 6 \\ -7 \end{pmatrix}$$

then $c_1 = c_2 = -1$ and $c_3 = 6$.

30. We have $\det(\mathbf{A} - \lambda\mathbf{I}) = (\lambda + 1)^2 = 0$. For $\lambda_1 = -1$ we obtain

$$\mathbf{K} = \begin{pmatrix} 1 \\ 1 \end{pmatrix}.$$

A solution of $(\mathbf{A} - \lambda_1\mathbf{I})\mathbf{P} = \mathbf{K}$ is

$$\mathbf{P} = \begin{pmatrix} 0 \\ 1/5 \end{pmatrix}$$

so that

$$\mathbf{X} = c_1 \begin{pmatrix} 1 \\ 1 \end{pmatrix} e^{-t} + c_2 \left[\begin{pmatrix} 1 \\ 1 \end{pmatrix} te^{-t} + \begin{pmatrix} 0 \\ 1/5 \end{pmatrix} e^{-t} \right].$$

117

33. We have $\det(\mathbf{A} - \lambda\mathbf{I}) = (1 - \lambda)(\lambda - 2)^2 = 0$. For $\lambda_1 = 1$ we obtain

$$\mathbf{K}_1 = \begin{pmatrix} 1 \\ 1 \\ 1 \end{pmatrix}.$$

For $\lambda_2 = 2$ we obtain

$$\mathbf{K}_2 = \begin{pmatrix} 1 \\ 0 \\ 1 \end{pmatrix} \quad \text{and} \quad \mathbf{K}_3 = \begin{pmatrix} 1 \\ 1 \\ 0 \end{pmatrix}.$$

Then

$$\mathbf{X} = c_1 \begin{pmatrix} 1 \\ 1 \\ 1 \end{pmatrix} e^t + c_2 \begin{pmatrix} 1 \\ 0 \\ 1 \end{pmatrix} e^{2t} + c_3 \begin{pmatrix} 1 \\ 1 \\ 0 \end{pmatrix} e^{2t}.$$

36. We have $\det(\mathbf{A} - \lambda\mathbf{I}) = (1 - \lambda)(\lambda - 2)^2 = 0$. For $\lambda_1 = 1$ we obtain

$$\mathbf{K}_1 = \begin{pmatrix} 1 \\ 0 \\ 0 \end{pmatrix}.$$

For $\lambda_2 = 2$ we obtain

$$\mathbf{K} = \begin{pmatrix} 0 \\ -1 \\ 1 \end{pmatrix}.$$

A solution of $(\mathbf{A} - \lambda_2\mathbf{I})\mathbf{P} = \mathbf{K}$ is

$$\mathbf{P} = \begin{pmatrix} 0 \\ -1 \\ 0 \end{pmatrix}$$

so that

$$\mathbf{X} = c_1 \begin{pmatrix} 1 \\ 0 \\ 0 \end{pmatrix} e^t + c_2 \begin{pmatrix} 0 \\ -1 \\ 1 \end{pmatrix} e^{2t} + c_3 \left[\begin{pmatrix} 0 \\ -1 \\ 1 \end{pmatrix} te^{2t} + \begin{pmatrix} 0 \\ -1 \\ 0 \end{pmatrix} e^{2t} \right].$$

39. We have $\det(\mathbf{A} - \lambda\mathbf{I}) = (\lambda - 4)^2 = 0$. For $\lambda_1 = 4$ we obtain

$$\mathbf{K} = \begin{pmatrix} 2 \\ 1 \end{pmatrix}.$$

A solution of $(\mathbf{A} - \lambda_1\mathbf{I})\mathbf{P} = \mathbf{K}$ is

$$\mathbf{P} = \begin{pmatrix} 1 \\ 1 \end{pmatrix}$$

so that

$$\mathbf{X} = c_1 \begin{pmatrix} 2 \\ 1 \end{pmatrix} e^{4t} + c_2 \left[\begin{pmatrix} 2 \\ 1 \end{pmatrix} t e^{4t} + \begin{pmatrix} 1 \\ 1 \end{pmatrix} e^{4t} \right].$$

If

$$\mathbf{X}(0) = \begin{pmatrix} -1 \\ 6 \end{pmatrix}$$

then $c_1 = -7$ and $c_2 = 13$.

42. We have $\det(\mathbf{A} - \lambda\mathbf{I}) = (\lambda + 3/25)(\lambda + 1/25) = 0$. For $\lambda_1 = -3/25$ and $\lambda_2 = -1/25$ we obtain

$$\mathbf{K}_1 = \begin{pmatrix} -1 \\ 2 \end{pmatrix} \quad \text{and} \quad \mathbf{K}_2 = \begin{pmatrix} 1 \\ 2 \end{pmatrix}$$

so that

$$\boldsymbol{\Phi}(t) = \begin{pmatrix} -e^{-3t/25} & e^{-t/25} \\ 2e^{-3t/25} & 2e^{-t/25} \end{pmatrix} \quad \text{and} \quad \boldsymbol{\Phi}^{-1}(t) = -\frac{1}{4} e^{4t/25} \begin{pmatrix} 2e^{-t/25} & e^{-t/25} \\ -2e^{-3t/25} & -e^{-3t/25} \end{pmatrix}.$$

Then

$$\mathbf{X} = \boldsymbol{\Phi}(t)\boldsymbol{\Phi}^{-1}(0)\mathbf{X}(0) = \begin{pmatrix} \frac{25}{2}e^{-3t/25} + \frac{25}{2}e^{-t/25} \\ -25e^{-3t/25} + 25e^{-t/25} \end{pmatrix}.$$

Exercises 8.7

3. Solving

$$\begin{vmatrix} 1 - \lambda & 3 \\ 3 & 1 - \lambda \end{vmatrix} = \lambda^2 - 2\lambda - 8 = (\lambda - 4)(\lambda + 2) = 0$$

we obtain eigenvalues $\lambda_1 = -2$ and $\lambda_2 = 4$. Corresponding eigenvectors are

$$\mathbf{K}_1 = \begin{pmatrix} 1 \\ -1 \end{pmatrix} \quad \text{and} \quad \mathbf{K}_2 = \begin{pmatrix} 1 \\ 1 \end{pmatrix}.$$

Thus

$$\mathbf{X}_c = c_1 \begin{pmatrix} 1 \\ -1 \end{pmatrix} e^{-2t} + c_2 \begin{pmatrix} 1 \\ 1 \end{pmatrix} e^{4t}.$$

Substituting

$$\mathbf{X}_p = \begin{pmatrix} a_3 \\ b_3 \end{pmatrix} t^2 + \begin{pmatrix} a_2 \\ b_2 \end{pmatrix} t + \begin{pmatrix} a_1 \\ b_1 \end{pmatrix}$$

into the system yields

$$a_3 + 3b_3 = 2 \qquad a_2 + 3b_2 = 2a_3 \qquad a_1 + 3b_1 = a_2$$

$$3a_3 + b_3 = 0 \qquad 3a_2 + b_2 + 1 = 2b_3 \qquad 3a_1 + b_1 + 5 = b_2$$

119

from which we obtain $a_3 = -1/4$, $b_3 = 3/4$, $a_2 = 1/4$, $b_2 = -1/4$, $a_1 = -2$, and $b_1 = 3/4$. Then

$$\mathbf{X}(t) = c_1 \begin{pmatrix} 1 \\ -1 \end{pmatrix} e^{-2t} + c_2 \begin{pmatrix} 1 \\ 1 \end{pmatrix} e^{4t} + \begin{pmatrix} -1/4 \\ 3/4 \end{pmatrix} t^2 + \begin{pmatrix} 1/4 \\ -1/4 \end{pmatrix} t + \begin{pmatrix} -2 \\ 3/4 \end{pmatrix}.$$

6. Solving

$$\begin{vmatrix} -1 - \lambda & 5 \\ -1 & 1 - \lambda \end{vmatrix} = \lambda^2 + 4 = 0$$

we obtain the eigenvalues $\lambda_1 = 2i$ and $\lambda_2 = -2i$. Corresponding eigenvectors are

$$\mathbf{K}_1 = \begin{pmatrix} 5 \\ 1 + 2i \end{pmatrix} \quad \text{and} \quad \mathbf{K}_2 = \begin{pmatrix} 5 \\ 1 - 2i \end{pmatrix}.$$

Thus

$$\mathbf{X}_c = c_1 \begin{pmatrix} 5 \cos 2t \\ \cos 2t - 2 \sin 2t \end{pmatrix} + c_2 \begin{pmatrix} 5 \sin 2t \\ 2 \cos 2t + \sin 2t \end{pmatrix}.$$

Substituting

$$\mathbf{X}_p = \begin{pmatrix} a_2 \\ b_2 \end{pmatrix} \cos t + \begin{pmatrix} a_1 \\ b_1 \end{pmatrix} \sin t$$

into the system yields

$$-a_2 + 5b_2 - a_1 = 0$$

$$-a_2 + b_2 - b_1 - 2 = 0$$

$$-a_1 + 5b_1 + a_2 + 1 = 0$$

$$-a_1 + b_1 + b_2 = 0$$

from which we obtain $a_2 = -3$, $b_2 = -2/3$, $a_1 = -1/3$, and $b_1 = 1/3$. Then

$$\mathbf{X}(t) = c_1 \begin{pmatrix} 5 \cos 2t \\ \cos 2t - 2 \sin 2t \end{pmatrix} + c_2 \begin{pmatrix} 5 \sin 2t \\ 2 \cos 2t + \sin 2t \end{pmatrix} + \begin{pmatrix} -3 \\ -2/3 \end{pmatrix} \cos t + \begin{pmatrix} -1/3 \\ 1/3 \end{pmatrix} \sin t.$$

9. Solving

$$\begin{vmatrix} -1 - \lambda & -2 \\ 3 & 4 - \lambda \end{vmatrix} = \lambda^2 - 3\lambda + 2 = (\lambda - 1)(\lambda - 2) = 0$$

we obtain the eigenvalues $\lambda_1 = 1$ and $\lambda_2 = 2$. Corresponding eigenvectors are

$$\mathbf{K}_1 = \begin{pmatrix} 1 \\ -1 \end{pmatrix} \quad \text{and} \quad \mathbf{K}_2 = \begin{pmatrix} -4 \\ 6 \end{pmatrix}.$$

Thus

$$\mathbf{X}_c = c_1 \begin{pmatrix} 1 \\ -1 \end{pmatrix} e^t + c_2 \begin{pmatrix} -4 \\ 6 \end{pmatrix} e^{2t}.$$

Substituting

$$\mathbf{X}_p = \begin{pmatrix} a_1 \\ b_1 \end{pmatrix}$$

into the system yields

$$-a_1 - 2b_1 = -3$$

$$3a_1 + 4b_1 = -3$$

from which we obtain $a_1 = -9$ and $b_1 = 6$. Then

$$\mathbf{X}(t) = c_1 \begin{pmatrix} 1 \\ -1 \end{pmatrix} e^t + c_2 \begin{pmatrix} -4 \\ 6 \end{pmatrix} e^{2t} + \begin{pmatrix} -9 \\ 6 \end{pmatrix}.$$

Setting

$$\mathbf{X}(0) = \begin{pmatrix} -4 \\ 5 \end{pmatrix}$$

we obtain

$$c_1 - 4c_2 - 9 = -4$$

$$-c_1 + 6c_2 + 6 = 5.$$

Then $c_1 = 13$ and $c_2 = 2$ so

$$\mathbf{X}(t) = 13 \begin{pmatrix} 1 \\ -1 \end{pmatrix} e^t + 2 \begin{pmatrix} -4 \\ 6 \end{pmatrix} e^{2t} + \begin{pmatrix} -9 \\ 6 \end{pmatrix}.$$

Exercises 8.8

3. From

$$\mathbf{X}' = \begin{pmatrix} 3 & -5 \\ 3/4 & -1 \end{pmatrix} \mathbf{X} + \begin{pmatrix} 1 \\ -1 \end{pmatrix} e^{t/2}$$

we obtain

$$\mathbf{X}_c = c_1 \begin{pmatrix} 10 \\ 3 \end{pmatrix} e^{3t/2} + c_2 \begin{pmatrix} 2 \\ 1 \end{pmatrix} e^{t/2}.$$

Then

$$\mathbf{\Phi} = \begin{pmatrix} 10e^{3t/2} & 2e^{t/2} \\ 3e^{3t/2} & e^{t/2} \end{pmatrix} \quad \text{and} \quad \mathbf{\Phi}^{-1} = \begin{pmatrix} \frac{1}{4}e^{-3t/2} & -\frac{1}{2}e^{-3t/2} \\ -\frac{3}{4}e^{-t/2} & \frac{5}{2}e^{-t/2} \end{pmatrix}$$

so that

$$\mathbf{U} = \int \mathbf{\Phi}^{-1}\mathbf{F}\,dt = \int \begin{pmatrix} \frac{3}{4}e^{-t} \\ -\frac{13}{4} \end{pmatrix} dt = \begin{pmatrix} -\frac{3}{4}e^{-t} \\ -\frac{13}{4}t \end{pmatrix}$$

and

$$\mathbf{X}_p = \mathbf{\Phi}\mathbf{U} = \begin{pmatrix} -13/2 \\ -13/4 \end{pmatrix} te^{t/2} + \begin{pmatrix} -15/2 \\ -9/4 \end{pmatrix} e^{t/2}.$$

6. From

$$\mathbf{X}' = \begin{pmatrix} 0 & 2 \\ -1 & 3 \end{pmatrix} \mathbf{X} + \begin{pmatrix} 2 \\ e^{-3t} \end{pmatrix}$$

we obtain

$$\mathbf{X}_c = c_1 \begin{pmatrix} 2 \\ 1 \end{pmatrix} e^t + c_2 \begin{pmatrix} 1 \\ 1 \end{pmatrix} e^{2t}.$$

Then

$$\mathbf{\Phi} = \begin{pmatrix} 2e^t & e^{2t} \\ e^t & e^{2t} \end{pmatrix} \quad \text{and} \quad \mathbf{\Phi}^{-1} = \begin{pmatrix} e^{-t} & -e^{-t} \\ -e^{-2t} & 2e^{-2t} \end{pmatrix}$$

so that

$$\mathbf{U} = \int \mathbf{\Phi}^{-1} \mathbf{F} \, dt = \int \begin{pmatrix} 2e^{-t} - e^{-4t} \\ -2e^{-2t} + 2e^{-5t} \end{pmatrix} dt = \begin{pmatrix} -2e^{-t} + \frac{1}{4}e^{-4t} \\ e^{-2t} - \frac{2}{5}e^{-5t} \end{pmatrix}$$

and

$$\mathbf{X}_p = \mathbf{\Phi}\mathbf{U} = \begin{pmatrix} \frac{1}{10}e^{-3t} - 3 \\ -\frac{3}{20}e^{-3t} - 1 \end{pmatrix}.$$

9. From

$$\mathbf{X}' = \begin{pmatrix} 3 & 2 \\ -2 & -1 \end{pmatrix} \mathbf{X} + \begin{pmatrix} 2 \\ 1 \end{pmatrix} e^{-t}$$

we obtain

$$\mathbf{X}_c = c_1 \begin{pmatrix} 1 \\ -1 \end{pmatrix} e^t + c_2 \left[\begin{pmatrix} 1 \\ -1 \end{pmatrix} te^t + \begin{pmatrix} 0 \\ 1/2 \end{pmatrix} e^t \right].$$

Then

$$\mathbf{\Phi} = \begin{pmatrix} e^t & te^t \\ -e^t & \frac{1}{2}e^t - te^t \end{pmatrix} \quad \text{and} \quad \mathbf{\Phi}^{-1} = \begin{pmatrix} e^{-t} - 2te^{-t} & -2te^{-t} \\ 2e^{-t} & 2e^{-t} \end{pmatrix}$$

so that

$$\mathbf{U} = \int \mathbf{\Phi}^{-1} \mathbf{F} \, dt = \int \begin{pmatrix} 2e^{-2t} - 6te^{-2t} \\ 6e^{-2t} \end{pmatrix} dt = \begin{pmatrix} \frac{1}{2}e^{-2t} + 3te^{-2t} \\ -3e^{-2t} \end{pmatrix}$$

and

$$\mathbf{X}_p = \mathbf{\Phi}\mathbf{U} = \begin{pmatrix} 1/2 \\ -2 \end{pmatrix} e^{-t}.$$

12. From

$$\mathbf{X}' = \begin{pmatrix} 1 & -1 \\ 1 & 1 \end{pmatrix} \mathbf{X} + \begin{pmatrix} 3 \\ 3 \end{pmatrix} e^t$$

we obtain

$$\mathbf{X}_c = c_1 \begin{pmatrix} -\sin t \\ \cos t \end{pmatrix} e^t + c_2 \begin{pmatrix} \cos t \\ \sin t \end{pmatrix} e^t.$$

Then

$$\mathbf{\Phi} = \begin{pmatrix} -\sin t & \cos t \\ \cos t & \sin t \end{pmatrix} e^t \quad \text{and} \quad \mathbf{\Phi}^{-1} = \begin{pmatrix} -\sin t & \cos t \\ \cos t & \sin t \end{pmatrix} e^{-t}.$$

so that

$$\mathbf{U} = \int \mathbf{\Phi}^{-1}\mathbf{F}\,dt = \int \begin{pmatrix} -3\sin t + 3\cos t \\ 3\cos t + 3\sin t \end{pmatrix} dt = \begin{pmatrix} 3\cos t + 3\sin t \\ 3\sin t - 3\cos t \end{pmatrix}$$

and

$$\mathbf{X}_p = \mathbf{\Phi}\mathbf{U} = \begin{pmatrix} -3 \\ 3 \end{pmatrix} e^t.$$

15. From

$$\mathbf{X}' = \begin{pmatrix} 0 & 1 \\ -1 & 0 \end{pmatrix} \mathbf{X} + \begin{pmatrix} 0 \\ \sec t \tan t \end{pmatrix}$$

we obtain

$$\mathbf{X}_c = c_1 \begin{pmatrix} \cos t \\ -\sin t \end{pmatrix} + c_2 \begin{pmatrix} \sin t \\ \cos t \end{pmatrix}.$$

Then

$$\mathbf{\Phi} = \begin{pmatrix} \cos t & \sin t \\ -\sin t & \cos t \end{pmatrix} t \quad \text{and} \quad \mathbf{\Phi}^{-1} = \begin{pmatrix} \cos t & -\sin t \\ \sin t & \cos t \end{pmatrix}$$

so that

$$\mathbf{U} = \int \mathbf{\Phi}^{-1}\mathbf{F}\,dt = \int \begin{pmatrix} -\tan^2 t \\ \tan t \end{pmatrix} dt = \begin{pmatrix} t - \tan t \\ \ln|\sec t| \end{pmatrix}$$

and

$$\mathbf{X}_p = \mathbf{\Phi}\mathbf{U} = \begin{pmatrix} \cos t \\ -\sin t \end{pmatrix} t + \begin{pmatrix} -\sin t \\ \sin t \tan t \end{pmatrix} + \begin{pmatrix} \sin t \\ \cos t \end{pmatrix} \ln|\sec t|.$$

18. From

$$\mathbf{X}' = \begin{pmatrix} 1 & -2 \\ 1 & -1 \end{pmatrix} \mathbf{X} + \begin{pmatrix} \tan t \\ 1 \end{pmatrix}$$

we obtain

$$\mathbf{X}_c = c_1 \begin{pmatrix} \cos t - \sin t \\ \cos t \end{pmatrix} + c_2 \begin{pmatrix} \cos t + \sin t \\ \sin t \end{pmatrix}.$$

Then

$$\mathbf{\Phi} = \begin{pmatrix} \cos t - \sin t & \cos t + \sin t \\ \cos t & \sin t \end{pmatrix} \quad \text{and} \quad \mathbf{\Phi}^{-1} = \begin{pmatrix} -\sin t & \cos t + \sin t \\ \cos t & \sin t - \cos t \end{pmatrix}$$

so that

$$\mathbf{U} = \int \mathbf{\Phi}^{-1}\mathbf{F}\,dt = \int \begin{pmatrix} 2\cos t + \sin t - \sec t \\ 2\sin t - \cos t \end{pmatrix} dt = \begin{pmatrix} 2\sin t - \cos t - \ln|\sec t + \tan t| \\ -2\cos t - \sin t \end{pmatrix}$$

and

$$\mathbf{X}_p = \mathbf{\Phi}\mathbf{U} = \begin{pmatrix} 3\sin t \cos t - \cos^2 t - 2\sin^2 t + (\sin t - \cos t)\ln|\sec t + \tan t| \\ \sin^2 t - \cos^2 t - \cos t(\ln|\sec t + \tan t|) \end{pmatrix}.$$

21. From

$$\mathbf{X}' = \begin{pmatrix} 3 & -1 \\ -1 & 3 \end{pmatrix} \mathbf{X} + \begin{pmatrix} 4e^{2t} \\ 4e^{4t} \end{pmatrix}$$

123

we obtain

$$\Phi = \begin{pmatrix} -e^{4t} & e^{2t} \\ e^{4t} & e^{2t} \end{pmatrix}, \quad \Phi^{-1} = \begin{pmatrix} -\frac{1}{2}e^{-4t} & \frac{1}{2}e^{4t} \\ \frac{1}{2}e^{-2t} & \frac{1}{2}e^{2t} \end{pmatrix},$$

and

$$X = \Phi\Phi^{-1}(0)X(0) + \Phi\int_0^t \Phi^{-1}F\,ds = \Phi \cdot \begin{pmatrix} 0 \\ 1 \end{pmatrix} + \Phi \cdot \begin{pmatrix} e^{-2t} + 2t - 1 \\ e^{2t} + 2t - 1 \end{pmatrix}$$

$$= \begin{pmatrix} 2 \\ 2 \end{pmatrix} te^{2t} + \begin{pmatrix} -1 \\ 1 \end{pmatrix} e^{2t} + \begin{pmatrix} -2 \\ 2 \end{pmatrix} te^{4t} + \begin{pmatrix} 2 \\ 0 \end{pmatrix} e^{4t}.$$

24. From

$$X' = \begin{pmatrix} 3 & -2 \\ 5 & -3 \end{pmatrix} X + \begin{pmatrix} 2 \\ 3 \end{pmatrix}$$

we obtain

$$\Psi = \begin{pmatrix} \sin t - 3\cos t & 2\cos t \\ -5\cos t & \sin t + 3\cos t \end{pmatrix}, \quad \Psi^{-1} = \begin{pmatrix} \sin t + 3\cos t & -2\cos t \\ 5\cos t & \sin t - 3\cos t \end{pmatrix},$$

and

$$X = \Psi X(\pi/2) + \Psi\int_{\pi/2}^t \Psi^{-1}F\,ds = \Psi \cdot \begin{pmatrix} 0 \\ 0 \end{pmatrix} + \Psi \cdot \begin{pmatrix} -2\cos t \\ \sin t - 3\cos t - 1 \end{pmatrix}$$

$$= \begin{pmatrix} 0 \\ 1 \end{pmatrix} - \begin{pmatrix} 2 \\ 3 \end{pmatrix}\cos t - \begin{pmatrix} 0 \\ 1 \end{pmatrix}\sin t.$$

Exercises 8.9

3. Using the result of Problem 1

$$X = \begin{pmatrix} \cosh t & \sinh t \\ \sinh t & \cosh t \end{pmatrix}\begin{pmatrix} c_1 \\ c_2 \end{pmatrix} = c_1 \begin{pmatrix} \cosh t \\ \sinh t \end{pmatrix} + c_2 \begin{pmatrix} \sinh t \\ \cosh t \end{pmatrix}.$$

6. To solve

$$X' = \begin{pmatrix} 0 & 1 \\ 1 & 0 \end{pmatrix} X + \begin{pmatrix} \cosh t \\ \sinh t \end{pmatrix}$$

we identify $t_0 = 0$, $F(s) = \begin{pmatrix} \cosh t \\ \sinh t \end{pmatrix}$, and use the results of Problem 1 and equation (3) in the text.

$$\mathbf{X}(t) = e^{t\mathbf{A}}\mathbf{C} + e^{t\mathbf{A}} \int_{t_0}^{t} e^{-s\mathbf{A}}\mathbf{F}(s)\, ds$$

$$= \begin{pmatrix} \cosh t & \sinh t \\ \sinh t & \cosh t \end{pmatrix} \begin{pmatrix} c_1 \\ c_2 \end{pmatrix} + \begin{pmatrix} \cosh t & \sinh t \\ \sinh t & \cosh t \end{pmatrix} \int_0^t \begin{pmatrix} \cosh s & -\sinh s \\ -\sinh s & \cosh s \end{pmatrix} \begin{pmatrix} \cosh s \\ \sinh s \end{pmatrix} ds$$

$$= \begin{pmatrix} c_1 \cosh t + c_2 \sinh t \\ c_1 \sinh t + c_2 \cosh t \end{pmatrix} + \begin{pmatrix} \cosh t & \sinh t \\ \sinh t & \cosh t \end{pmatrix} \int_0^t \begin{pmatrix} 1 \\ 0 \end{pmatrix} ds$$

$$= \begin{pmatrix} c_1 \cosh t + c_2 \sinh t \\ c_1 \sinh t + c_2 \cosh t \end{pmatrix} + \begin{pmatrix} \cosh t & \sinh t \\ \sinh t & \cosh t \end{pmatrix} \begin{pmatrix} s \\ 0 \end{pmatrix} \Big|_0^t$$

$$= \begin{pmatrix} c_1 \cosh t + c_2 \sinh t \\ c_1 \sinh t + c_2 \cosh t \end{pmatrix} + \begin{pmatrix} \cosh t & \sinh t \\ \sinh t & \cosh t \end{pmatrix} \begin{pmatrix} t \\ 0 \end{pmatrix}$$

$$= \begin{pmatrix} c_1 \cosh t + c_2 \sinh t \\ c_1 \sinh t + c_2 \cosh t \end{pmatrix} + \begin{pmatrix} t \cosh t \\ t \sinh t \end{pmatrix} = c_1 \begin{pmatrix} \cosh t \\ \sinh t \end{pmatrix} + c_2 \begin{pmatrix} \sinh t \\ \cosh t \end{pmatrix} + t \begin{pmatrix} \cosh t \\ \sinh t \end{pmatrix}.$$

9. Solving

$$\begin{vmatrix} 2 - \lambda & 1 \\ -3 & 6 - \lambda \end{vmatrix} = \lambda^2 - 8\lambda + 15 = (\lambda - 3)(\lambda - 5) = 0$$

we find eigenvalues $\lambda_1 = 3$ and $\lambda_2 = 5$. Corresponding eigenvectors are

$$\mathbf{K}_1 = \begin{pmatrix} 1 \\ 1 \end{pmatrix} \quad \text{and} \quad \mathbf{K}_2 = \begin{pmatrix} 1 \\ 3 \end{pmatrix}.$$

Then

$$\mathbf{P} = \begin{pmatrix} 1 & 1 \\ 1 & 3 \end{pmatrix}, \quad \mathbf{P}^{-1} = \begin{pmatrix} 3/2 & -1/2 \\ -1/2 & 1/2 \end{pmatrix}, \quad \text{and} \quad \mathbf{D} = \begin{pmatrix} 3 & 0 \\ 0 & 5 \end{pmatrix},$$

so

$$\mathbf{PDP}^{-1} = \begin{pmatrix} 2 & 1 \\ -3 & 6 \end{pmatrix}.$$

12. From equation (2) in the text

125

$$e^{t\mathbf{D}} = \begin{pmatrix} 1 & 0 & \cdots & 0 \\ 0 & 1 & \cdots & 0 \\ \vdots & \vdots & \ddots & \vdots \\ 0 & 0 & \cdots & 1 \end{pmatrix} + \begin{pmatrix} \lambda_1 & 0 & \cdots & 0 \\ 0 & \lambda_2 & \cdots & 0 \\ \vdots & \vdots & \ddots & \vdots \\ 0 & 0 & \cdots & \lambda_n \end{pmatrix} + \frac{1}{2!}t^2 \begin{pmatrix} \lambda_1^2 & 0 & \cdots & 0 \\ 0 & \lambda_2^2 & \cdots & 0 \\ \vdots & \vdots & \ddots & \vdots \\ 0 & 0 & \cdots & \lambda_n^2 \end{pmatrix}$$

$$+ \frac{1}{3!}t^3 \begin{pmatrix} \lambda_1^3 & 0 & \cdots & 0 \\ 0 & \lambda_2^3 & \cdots & 0 \\ \vdots & \vdots & \ddots & \vdots \\ 0 & 0 & \cdots & \lambda_n^3 \end{pmatrix} + \cdots$$

$$= \begin{pmatrix} 1 + \lambda_1 t + \frac{1}{2!}(\lambda_1 t)^2 + \cdots & 0 & \cdots & 0 \\ 0 & 1 + \lambda_2 t + \frac{1}{2!}(\lambda_2 t)^2 + \cdots & \cdots & 0 \\ \vdots & \vdots & \ddots & \vdots \\ 0 & 0 & \cdots & 1 + \lambda_n t + \frac{1}{2!}(\lambda_n t)^2 + \cdots \end{pmatrix}$$

$$= \begin{pmatrix} e^{\lambda_1 t} & 0 & \cdots & 0 \\ 0 & e^{\lambda_2 t} & \cdots & 0 \\ \vdots & \vdots & \ddots & \vdots \\ 0 & 0 & \cdots & e^{\lambda_n t} \end{pmatrix}$$

—— Chapter 8 Review Exercises ——

3. $\mathbf{A}^{-1} = -\frac{1}{2}\begin{pmatrix} 4 & -2 \\ -3 & 1 \end{pmatrix} = \begin{pmatrix} -2 & 1 \\ 3/2 & -1/2 \end{pmatrix}$

6. True, by Theorem 8.8.

9. True, by the definition of an eigenvector.

12. False;

$$\begin{pmatrix} 1 & 1 & 1 & \Big| & 2 \\ 0 & 1 & 0 & \Big| & 3 \\ 0 & 0 & 0 & \Big| & 0 \end{pmatrix} \Longrightarrow \begin{pmatrix} 1 & 0 & 1 & \Big| & -1 \\ 0 & 1 & 0 & \Big| & 3 \\ 0 & 0 & 0 & \Big| & 0 \end{pmatrix}.$$

15. From $(D-2)x - y = -e^t$ and $-3x + (D-4)y = -7e^t$ we obtain $(D-1)(D-5)x = -4e^t$ so that

$$x = c_1 e^t + c_2 e^{5t} + te^t.$$

Then

$$y = (D-2)x + e^t = -c_1 e^t + 3c_2 e^{5t} - te^t + 2e^t.$$

18. Taking the Laplace transform of the system gives

$$s^2 \mathcal{L}\{x\} + s^2 \mathcal{L}\{y\} = \frac{1}{s-2}$$

$$2s \mathcal{L}\{x\} + s^2 \mathcal{L}\{y\} = -\frac{1}{s-2}$$

so that

$$\mathcal{L}\{x\} = \frac{2}{s(s-2)^2} = \frac{1}{2}\frac{1}{s} - \frac{1}{2}\frac{1}{s-2} + \frac{1}{(s-2)^2}$$

and

$$\mathcal{L}\{y\} = \frac{-s-2}{s^2(s-2)^2} = -\frac{3}{4}\frac{1}{s} - \frac{1}{2}\frac{1}{s^2} + \frac{3}{4}\frac{1}{s-2} - \frac{1}{(s-2)^2}.$$

Then

$$x = \frac{1}{2} - \frac{1}{2}e^{2t} + te^{2t} \quad \text{and} \quad y = -\frac{3}{4} - \frac{1}{2}t + \frac{3}{4}e^{2t} - te^{2t}.$$

21. Let $x_1 = x$, $x_2 = y$, $x_3 = Dx$, and $x_4 = Dy$ so that

$$Dx_1 = x_3$$

$$Dx_2 = x_4$$

$$Dx_3 = x_4 - 2x_3 - 2x_1 - \ln t + 10t - 4$$

$$Dx_4 = -x_3 - x_1 + 5t - 2.$$

24. We have $\det(\mathbf{A} - \lambda \mathbf{I}) = (\lambda + 6)(\lambda + 2) = 0$ so that

$$\mathbf{X} = c_1 \begin{pmatrix} 1 \\ -1 \end{pmatrix} e^{-6t} + c_2 \begin{pmatrix} 1 \\ 1 \end{pmatrix} e^{-2t}.$$

27. We have $\det(\mathbf{A} - \lambda \mathbf{I}) = \lambda^2(3 - \lambda) = 0$ so that

$$\mathbf{X} = c_1 \begin{pmatrix} -1 \\ 1 \\ 0 \end{pmatrix} + c_2 \begin{pmatrix} -1 \\ 0 \\ 1 \end{pmatrix} + c_3 \begin{pmatrix} 1 \\ 1 \\ 1 \end{pmatrix} e^{3t}.$$

30. We have

$$\mathbf{X}_c = c_1 \begin{pmatrix} 2\cos t \\ -\sin t \end{pmatrix} e^t + c_2 \begin{pmatrix} 2\sin t \\ \cos t \end{pmatrix} e^t.$$

Then

$$\boldsymbol{\Phi} = \begin{pmatrix} 2\cos t & 2\sin t \\ -\sin t & \cos t \end{pmatrix} e^t, \quad \boldsymbol{\Phi}^{-1} = \begin{pmatrix} \frac{1}{2}\cos t & -\sin t \\ \frac{1}{2}\sin t & \cos t \end{pmatrix} e^{-t},$$

and

$$\mathbf{U} = \int \boldsymbol{\Phi}^{-1}\mathbf{F}\,dt = \int \begin{pmatrix} \cos t - \sec t \\ \sin t \end{pmatrix} dt = \begin{pmatrix} \sin t - \ln|\sec t + \tan t| \\ -\cos t \end{pmatrix},$$

so that

$$\mathbf{X}_p = \boldsymbol{\Phi}\mathbf{U} = \begin{pmatrix} -2\cos t\ln|\sec t + \tan t| \\ -1 + \sin t\ln|\sec t + \tan t| \end{pmatrix}.$$

9 Numerical Methods for
Ordinary Differential Equations

Exercises 9.1

3.

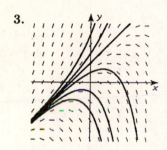

6. Setting $2x + y = c$ we obtain $y = -2x + c$; a family of lines with slope -2.

9. Setting $\sqrt{x^2 + y^2 + 2y + 1} = c$ we obtain $x^2 + (y+1)^2 = c^2$; a family of circles centered at $(0, -1)$.

12. Setting $y + e^x = c$ we obtain $y = c - e^x$; a family of exponential curves.

15. Setting $x = c$ we see that the isoclines form a family of vertical lines.

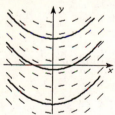

18. Setting $1/y = c$ we obtain the isoclines $y = 1/c$.

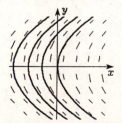

21. Setting $y - \cos\frac{\pi}{2}x = c$ we obtain the isoclines $y = \cos\frac{\pi}{2}x + c$.

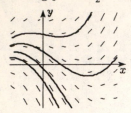

24. $y = cx$ is a solution of the differential equation if and only if

$$y' = c = \frac{\alpha x + \beta cx}{\gamma x + \delta cx}$$

if and only if

$$[\delta c^2 + (\gamma - \beta)c - \alpha]x = 0$$

if and only if

$$\delta c^2 + (\gamma - \beta)c - \alpha = 0$$

if and only if

$$c = \frac{\beta - \gamma \pm \sqrt{(\beta - \gamma)^2 + 4\alpha\delta}}{2\delta}$$

if and only if

$$(\beta - \gamma)^2 + 4\alpha\delta \geq 0.$$

27. The isoclines of $y' = 2x/y$ are $2x/y = c$ or

$$y = \frac{2}{c}x.$$

Setting $2/c = c$ we obtain $c = \pm\sqrt{2}$. Thus $y = \pm\sqrt{2}\,x$ are solutions of the differential equation .

30. The isoclines of $y' = (5x + 10y)/(-4x + 3y)$ are $(5x + 10y)/(-4x + 3y) = c$ or

$$y = \frac{4c + 5}{3c - 10}x.$$

Setting $(4c + 5)/(3c - 10) = c$ we obtain $c = 3c^2 - 14c - 5 = (3c + 1)(c - 5) = 0$. Thus $c = -1/3$ and $c = 5$ and $y = -\frac{1}{3}x$ and $y = 5x$ are solutions of the differential equation .

Exercises 9.2

All tables in this chapter were constructed in a spreadsheet program which does not support subscripts. Consequently, x_n and y_n will be indicated as $x(n)$ and $y(n)$, respectively.

3.

$h = 0.1$	
$x(n)$	$y(n)$
1.00	5.0000
1.10	3.8000
1.20	2.9800
1.30	2.4260
1.40	2.0582
1.50	1.8207

$h = 0.05$	
$x(n)$	$y(n)$
1.00	5.0000
1.05	4.4000
1.10	3.8950
1.15	3.4708
1.20	3.1151
1.25	2.8179
1.30	2.5702
1.35	2.3647
1.40	2.1950
1.45	2.0557
1.50	1.9424

6.

$h = 0.1$	
$x(n)$	$y(n)$
0.00	1.0000
0.10	1.1000
0.20	1.2220
0.30	1.3753
0.40	1.5735
0.50	1.8371

$h = 0.05$	
$x(n)$	$y(n)$
0.00	1.0000
0.05	1.0500
0.10	1.1053
0.15	1.1668
0.20	1.2360
0.25	1.3144
0.30	1.4039
0.35	1.5070
0.40	1.6267
0.45	1.7670
0.50	1.9332

9.

$h = 0.1$	
$x(n)$	$y(n)$
0.00	0.5000
0.10	0.5250
0.20	0.5431
0.30	0.5548
0.40	0.5613
0.50	0.5639

$h = 0.05$	
$x(n)$	$y(n)$
0.00	0.5000
0.05	0.5125
0.10	0.5232
0.15	0.5322
0.20	0.5395
0.25	0.5452
0.30	0.5496
0.35	0.5527
0.40	0.5547
0.45	0.5559
0.50	0.5565

Exercises 9.2

12.

h = 0.1	
x(n)	*y(n)*
1.00	0.5000
1.10	0.5250
1.20	0.5499
1.30	0.5747
1.40	0.5991
1.50	0.6231

h = 0.05	
x(n)	*y(n)*
1.00	0.5000
1.05	0.5125
1.10	0.5250
1.15	0.5375
1.20	0.5499
1.25	0.5623
1.30	0.5746
1.35	0.5868
1.40	0.5989
1.45	0.6109
1.50	0.6228

15.

h=0.1	EULER	IMPROVED EULER
x(n)	*y(n)*	*y(n)*
1.00	1.0000	1.0000
1.10	1.2000	1.2469
1.20	1.4938	1.6668
1.30	1.9711	2.6427
1.40	2.9060	8.7988

Exercises 9.3

3. We use

$$y'' = 2yy'$$

$$= 2y(1 + y^2)$$

$$= 2y + 2y^3$$

so that

$$y_{n+1} = y_n + (1 + y_n^2)h + (2y_n + 2y_n^3)\frac{1}{2}h^2.$$

h = 0.1	
x(n)	*y(n)*
0.00	0.0000
0.10	0.1000
0.20	0.2020
0.30	0.3082
0.40	0.4211
0.50	0.5438

h = 0.05	
x(n)	*y(n)*
0.00	0.0000
0.05	0.0500
0.10	0.1003
0.15	0.1510
0.20	0.2025
0.25	0.2551
0.30	0.3090
0.35	0.3647
0.40	0.4223
0.45	0.4825
0.50	0.5456

6. We use

$$y'' = 1 + 2yy'$$

$$= 1 + 2y(x + y^2)$$

$$= 1 + 2xy + 2y^3$$

so that

$$y_{n+1} = y_n + (x_n + y_n^2)h + (1 + 2x_ny_n + 2y_n^3)\frac{1}{2}h^2.$$

h = 0.1	
x(n)	y(n)
0.00	0.0000
0.10	0.0050
0.20	0.0200
0.30	0.0451
0.40	0.0804
0.50	0.1264

h = 0.05	
x(n)	y(n)
0.00	0.0000
0.05	0.0013
0.10	0.0050
0.15	0.0113
0.20	0.0200
0.25	0.0313
0.30	0.0451
0.35	0.0615
0.40	0.0805
0.45	0.1021
0.50	0.1265

9. We use

$$y'' = 2xyy' + y^2 - \frac{xy' - y}{x^2}$$

$$= 2xy\left(xy^2 - \frac{y}{x}\right) + y^2$$

$$- \frac{1}{x}\left(xy^2 - \frac{y}{x}\right) + \frac{y}{x^2}$$

$$= 2x^2y^3 - 2y^2 + \frac{2y}{x^2}$$

so that

$$y_{n+1} = y_n + \left(x_ny_n^2 - \frac{y_n}{x_n}\right)h + \left(2x_n^2y_n^3 - 2y_n^2 + \frac{2y_n}{x_n^2}\right)\frac{1}{2}h^2.$$

h = 0.1	
x(n)	y(n)
1.00	1.0000
1.10	1.0100
1.20	1.0410
1.30	1.0969
1.40	1.1857
1.50	1.3226

h = 0.05	
x(n)	y(n)
1.00	1.0000
1.05	1.0025
1.10	1.0101
1.15	1.0229
1.20	1.0415
1.25	1.0663
1.30	1.0983
1.35	1.1387
1.40	1.1891
1.45	1.2518
1.50	1.3301

12. Let $f(x, y) = \alpha x + \beta y$ so that $f_x = \alpha, f_y = \beta$, and all higher derivatives are 0. Using the Taylor series expansion for $f(x, y)$ we have

$$f(x_{n+1}, y_{n+1}^*) = f(x_n + h, y_n + hf(x_n, y_n))$$

$$= f(x_n, y_n) + f_x(x_n, y_n)h + f_y(x_n, y_n)hf(x_n, y_n)$$

$$= f(x_n, y_n) + \alpha h + \beta hf(x_n, y_n).$$

Since $f(x_n, y_n) = y_n'$ and $\alpha + \beta y_n' = y_n''$ we have

$$y_{n+1} = y_n + \frac{1}{2}h[f(x_n, y_n) + f(x_{n+1}, y_{n+1}^*)]$$

$$= y_n + \frac{1}{2}h[f(x_n, y_n) + f(x_n, y_n) + \alpha h + \beta hf(x_n, y_n)]$$

$$= y_n + \frac{1}{2}h[2y_n' + h(\alpha + \beta y_n')]$$

$$= y_n + hy_n' + \frac{1}{2}h^2y_n''.$$

Exercises 9.4

3.

x(n)	y(n)
0.00	0.0000
0.10	0.1003
0.20	0.2027
0.30	0.3093
0.40	0.4228
0.50	0.5463

6.

x(n)	y(n)
0.00	0.0000
0.10	0.0050
0.20	0.0200
0.30	0.0451
0.40	0.0805
0.50	0.1266

9.

x(n)	y(n)
1.00	1.0000
1.10	1.0101
1.20	1.0417
1.30	1.0989
1.40	1.1905
1.50	1.3333

12. Separating variables and using partial fractions we have

$$\frac{1}{2\sqrt{32}}\left(\frac{1}{\sqrt{32}-\sqrt{0.025}\,v}+\frac{1}{\sqrt{32}+\sqrt{0.025}\,v}\right)dv=dt$$

and

$$\frac{1}{2\sqrt{32}\sqrt{0.025}}\left(\ln\left|\sqrt{32}+\sqrt{0.025}\,v\right|-\ln\left|\sqrt{32}-\sqrt{0.025}\,v\right|\right)=t+c.$$

Since $v(0)=0$ we find $c=0$. Solving for v we obtain

$$v(t)=\frac{16\sqrt{5}\left(e^{\sqrt{3.2}\,t}-1\right)}{e^{\sqrt{3.2}\,t}+1}$$

and $v(5)\approx 35.7678$.

15.

x(n)	y(n)
1.00	1.0000
1.10	1.2511
1.20	1.6934
1.30	2.9425
1.40	903.0282

Exercises 9.5

3.

x(n)	y(n)	
0.00	1.0000	initial condition
0.20	0.7328	Runge–Kutta
0.40	0.6461	Runge–Kutta
0.60	0.6585	Runge–Kutta
	0.7332	*predictor*
0.80	0.7232	corrector

6.

x(n)	y(n)	
0.00	1.0000	initial condition
0.20	1.4414	Runge–Kutta
0.40	1.9719	Runge–Kutta
0.60	2.6028	Runge–Kutta
	3.3483	*predictor*
0.80	3.3486	corrector
	4.2276	*predictor*
1.00	4.2280	corrector

x(n)	y(n)	
0.00	1.0000	initial condition
0.10	1.2102	Runge–Kutta
0.20	1.4414	Runge–Kutta
0.30	1.6949	Runge–Kutta
	1.9719	*predictor*
0.40	1.9719	corrector
	2.2740	*predictor*
0.50	2.2740	corrector
	2.6028	*predictor*
0.60	2.6028	corrector
	2.9603	*predictor*
0.70	2.9603	corrector
	3.3486	*predictor*
0.80	3.3486	corrector
	3.7703	*predictor*
0.90	3.7703	corrector
	4.2280	*predictor*
1.00	4.2280	corrector

9.

x(n)	y(n)	
0.00	1.0000	initial condition
0.10	1.0052	Runge–Kutta
0.20	1.0214	Runge–Kutta
0.30	1.0499	Runge–Kutta
	1.0918	predictor
0.40	1.0918	corrector

Exercises 9.6

3. Since the fourth-order Runge-Kutta formula agrees with the Taylor polynomial through $k = 4$, the local truncation error is

$$y^{(5)}(c)\,\frac{h^5}{5!}\quad\text{where}\quad x_n < c < x_{n+1}.$$

6. (a) Using the three-term Taylor method we obtain $y(0.1) \approx y_1 = 1.22$.

(b) Using $y''' = 8e^{2x}$ we see that the local truncation error is

$$y'''(c)\,\frac{h^3}{6} = 8e^{2c}\,\frac{(0.1)^3}{6} = 0.001333e^{2c}.$$

Since e^{2x} is an increasing function, $e^{2c} \le e^{2(0.1)} = e^{0.2}$ for $0 \le c \le 0.1$. Thus an upper bound for the local truncation error is $0.001333e^{0.2} = 0.001628$.

(c) Since $y(0.1) = e^{0.2} = 1.221403$, the actual error is $y(0.1) - y_1 = 0.001403$ which is less than 0.001628.

(d) Using the three-term Taylor method with $h = 0.05$ we obtain $y(0.1) \approx y_2 = 1.221025$.

(e) The error in **(d)** is $1.221403 - 1.221025 = 0.000378$. With global truncation error $O(h^2)$, when the step size is halved we expect the error for $h = 0.05$ to be one-fourth the error for $h = 0.1$. Comparing 0.000378 with 0.001403 we see that this is the case.

9. (a) Using the improved Euler method we obtain $y(0.1) \approx y_1 = 0.825$.

(b) Using $y''' = -10e^{-2x}$ we see that the local truncation error is

$$10e^{-2c}\,\frac{(0.1)^3}{6} = 0.001667e^{-2c}.$$

Since e^{-2x} is a decreasing function, $e^{-2c} \le e^0 = 1$ for $0 \le c \le 0.1$. Thus an upper bound for the local truncation error is $0.001667(1) = 0.001667$.

(c) Since $y(0.1) = 0.823413$, the actual error is $y(0.1) - y_1 = 0.001587$, which is less than 0.001667.

(d) Using the improved Euler method with $h = 0.05$ we obtain $y(0.1) \approx y_2 = 0.823781$.

(e) The error in **(d)** is $|0.823413 - 0.8237181| = 0.000305$. With global truncation error $O(h^2)$, when the step size is halved we expect the error for $h = 0.05$ to be one-fourth the error when $h = 0.1$. Comparing 0.000305 with 0.001587 we see that this is the case.

12. (a) Using $y'' = 38e^{-3(x-1)}$ we see that the local truncation error is

$$y''(c)\,\frac{h^2}{2} = 38e^{-3(c-1)}\,\frac{h^2}{2} = 19h^2e^{-3(c-1)}.$$

(b) Since $e^{-3(x-1)}$ is a decreasing function for $1 \le x \le 1.5$, $e^{-3(c-1)} \le e^{-3(1-1)} = 1$ for $1 \le c \le 1.5$ and

$$y''(c)\frac{h^2}{2} \le 19(0.1)^2(1) = 0.19.$$

(c) Using the Euler method with $h = 0.1$ we obtain $y(1.5) \approx 1.8207$. With $h = 0.05$ we obtain $y(1.5) \approx 1.9424$.

(d) Since $y(1.5) = 2.0532$, the error for $h = 0.1$ is $E_{0.1} = 0.2325$, while the error for $h = 0.05$ is $E_{0.05} = 0.1109$. With global truncation error $O(h)$ we expect $E_{0.1}/E_{0.05} \approx 2$. We actually have $E_{0.1}/E_{0.05} = 2.10$.

15. (a) Using $y^{(5)} = -1026e^{-3(x-1)}$ we see that the local truncation error is

$$\left| y^{(5)}(c)\frac{h^5}{120} \right| = 8.55h^5 e^{-3(c-1)}.$$

(b) Since $e^{-3(x-1)}$ is a decreasing function for $1 \le x \le 1.5$, $e^{-3(c-1)} \le e^{-3(1-1)} = 1$ for $1 \le c \le 1.5$ and

$$y^{(5)}(c)\frac{h^5}{120} \le 8.55(0.1)^5(1) = 0.0000855.$$

(c) Using the fourth-order Runge-Kutta method with $h = 0.1$ we obtain $y(1.5) \approx 2.053338827$. With $h = 0.05$ we obtain $y(1.5) \approx 2.053222989$.

(d) Since $y(1.5) = 2.053216232$, the error for $h = 0.1$ is $E_{0.1} = 0.000122595$, while the error for $h = 0.05$ is $E_{0.05} = 0.000006757$. With global truncation error $O(h^4)$ we expect $E_{0.1}/E_{0.05} \approx 16$. We actually have $E_{0.1}/E_{0.05} = 18.14$.

18. (a) Using $y''' = \dfrac{2}{(x+1)^3}$ we see that the local truncation error is

$$y'''(c)\frac{h^3}{6} = \frac{1}{(c+1)^3}\frac{h^3}{3}.$$

(b) Since $\dfrac{1}{(x+1)^3}$ is a decreasing function for $0 \le x \le 0.5$, $\dfrac{1}{(c+1)^3} \le \dfrac{1}{(0+1)^3} = 1$ for $0 \le c \le 0.5$ and

$$y'''(c)\frac{h^3}{6} \le (1)\frac{(0.1)^3}{3} = 0.000333.$$

(c) Using the three-term Taylor method with $h = 0.1$ we obtain $y(0.5) \approx 0.404643$. With $h = 0.05$ we obtain $y(0.5) \approx 0.405270$.

(d) Since $y(0.5) = 0.405465$, the error for $h = 0.1$ is $E_{0.1} = 0.000823$, while the error for $h = 0.05$ is $E_{0.05} = 0.000195$. With global truncation error $O(h^2)$ we expect $E_{0.1}/E_{0.05} \approx 4$. We actually have $E_{0.1}/E_{0.05} = 4.22$.

Exercises 9.7

3. The substitution $y' = u$ leads to the system

$$y' = u, \qquad u' = 4u - 4y.$$

Using formulas (5) and (6) in the text with x corresponding to t, y corresponding to x, and u corresponding to y, we obtain

Runge-Kutta method with h=0.2

m1	m2	m3	m4	k1	k2	k3	k4	x	y	u
								0.00	-2.0000	1.0000
0.2000	0.4400	0.5280	0.9072	2.4000	3.2800	3.5360	4.8064	0.20	-1.4928	4.4731

Runge-Kutta method with h=0.1

m1	m2	m3	m4	k1	k2	k3	k4	x	y	u
								0.00	-2.0000	1.0000
0.1000	0.1600	0.1710	0.2452	1.2000	1.4200	1.4520	1.7124	0.10	-1.8321	2.4427
0.2443	0.3298	0.3444	0.4487	1.7099	2.0031	2.0446	2.3900	0.20	-1.4919	4.4753

6.

Runge-Kutta method with h=0.1

m1	m2	m3	m4	k1	k2	k3	k4	t	i1	i2
								0.00	0.0000	0.0000
10.0000	0.0000	12.5000	-20.0000	0.0000	5.0000	-5.0000	22.5000	0.10	2.5000	3.7500
8.7500	-2.5000	13.4375	-28.7500	-5.0000	4.3750	-10.6250	29.6875	0.20	2.8125	5.7813
10.1563	-4.3750	17.0703	-40.0000	-8.7500	5.0781	-16.0156	40.3516	0.30	2.0703	7.4023
13.2617	-6.3672	22.9443	-55.1758	-12.7344	6.6309	-22.5488	55.3076	0.40	0.6104	9.1919
17.9712	-8.8867	31.3507	-75.9326	-17.7734	8.9856	-31.2024	75.9821	0.50	-1.5619	11.4877

9.

Runge-Kutta method with h=0.2

m1	m2	m3	m4	k1	k2	k3	k4	t	x	y
								0.00	-3.0000	5.0000
-1.0000	-0.9200	-0.9080	-0.8176	-0.6000	-0.7200	-0.7120	-0.8216	0.20	-3.9123	4.2857

Runge-Kutta method with h=0.1

m1	m2	m3	m4	k1	k2	k3	k4	t	x	y
								0.00	-3.0000	5.0000
-0.5000	-0.4800	-0.4785	-0.4571	-0.3000	-0.3300	-0.3290	-0.3579	0.10	-3.4790	4.6707
-0.4571	-0.4342	-0.4328	-0.4086	-0.3579	-0.3858	-0.3846	-0.4112	0.20	-3.9123	4.2857

3. We identify $P(x) = 2$, $Q(x) = 1$, $f(x) = 5x$, and $h = (1-0)/5 = 0.2$. Then the finite difference equation is

$$1.2y_{i+1} - 1.96y_i + 0.8y_{i-1} = 0.04(5x_i).$$

The solution of the corresponding linear system gives

x	0.0	0.2	0.4	0.6	0.8	1.0
y	0.0000	-0.2259	-0.3356	-0.3308	-0.2167	0.0000

6. We identify $P(x) = 5$, $Q(x) = 0$, $f(x) = 4\sqrt{x}$, and $h = (2-1)/6 = 0.1667$. Then the finite difference equation is

$$1.4167y_{i+1} - 2y_i + 0.5833y_{i-1} = 0.2778(4\sqrt{x_i}).$$

The solution of the corresponding linear system gives

x	1.0000	1.1667	1.3333	1.5000	1.6667	1.8333	2.0000
y	1.0000	-0.5918	-1.1626	-1.3070	-1.2704	-1.1541	-1.0000

9. We identify $P(x) = 1-x$, $Q(x) = x$, $f(x) = x$, and $h = (1-0)/10 = 0.1$. Then the finite difference equation is

$$[1 + 0.05(1 - x_i)]y_{i+1} + [-2 + 0.01x_i]y_i + [1 - 0.05(1 - x_i)]y_{i-1} = 0.01x_i.$$

The solution of the corresponding linear system gives

x	0.0	0.1	0.2	0.3	0.4	0.5	0.6
y	0.0000	0.2660	0.5097	0.7357	0.9471	1.1465	1.3353

0.7	0.8	0.9	1.0
1.5149	1.6855	1.8474	2.0000

12. We identify $P(r) = 2/r$, $Q(r) = 0$, $f(r) = 0$, and $h = (4-1)/6 = 0.5$. Then the finite difference equation is

$$\left(1 + \frac{0.5}{r_i}\right)u_{i+1} - 2u_i + \left(1 - \frac{0.5}{r_i}\right)u_{i-1} = 0.$$

The solution of the corresponding linear system gives

r	1.0	1.5	2.0	2.5	3.0	3.5	4.0
u	50.0000	72.2222	83.3333	90.0000	94.4444	97.6190	100.0000

Chapter 9 Review Exercises

3.

h=0.1 x(n)	EULER	IMPROVED EULER	3-TERM TAYLOR	RUNGE KUTTA
1.00	2.0000	2.0000	2.0000	2.0000
1.10	2.1386	2.1549	2.1556	2.1556
1.20	2.3097	2.3439	2.3446	2.3454
1.30	2.5136	2.5672	2.5680	2.5695
1.40	2.7504	2.8246	2.8255	2.8278
1.50	3.0201	3.1157	3.1167	3.1197

h=0.05 x(n)	EULER	IMPROVED EULER	3-TERM TAYLOR	RUNGE KUTTA
1.00	2.0000	2.0000	2.0000	2.0000
1.05	2.0693	2.0735	2.0735	2.0736
1.10	2.1469	2.1554	2.1555	2.1556
1.15	2.2328	2.2459	2.2460	2.2462
1.20	2.3272	2.3450	2.3451	2.3454
1.25	2.4299	2.4527	2.4528	2.4532
1.30	2.5409	2.5689	2.5690	2.5695
1.35	2.6604	2.6937	2.6938	2.6944
1.40	2.7883	2.8269	2.8271	2.8278
1.45	2.9245	2.9686	2.9688	2.9696
1.50	3.0690	3.1187	3.1188	3.1197

6.

h=0.1 x(n)	EULER	IMPROVED EULER	3-TERM TAYLOR	RUNGE KUTTA
1.00	1.0000	1.0000	1.0000	1.0000
1.10	1.2000	1.2380	1.2350	1.2415
1.20	1.4760	1.5910	1.5866	1.6036
1.30	1.8710	2.1524	2.1453	2.1909
1.40	2.4643	3.1458	3.1329	3.2745
1.50	3.4165	5.2510	5.2208	5.8338

h=0.05 x(n)	EULER	IMPROVED EULER	3-TERM TAYLOR	RUNGE KUTTA
1.00	1.0000	1.0000	1.0000	1.0000
1.05	1.1000	1.1091	1.1088	1.1095
1.10	1.2183	1.2405	1.2401	1.2415
1.15	1.3595	1.4010	1.4004	1.4029
1.20	1.5300	1.6001	1.5994	1.6036
1.25	1.7389	1.8523	1.8515	1.8586
1.30	1.9988	2.1799	2.1789	2.1911
1.35	2.3284	2.6197	2.6182	2.6401
1.40	2.7567	3.2360	3.2340	3.2755
1.45	3.3296	4.1528	4.1497	4.2363
1.50	4.1253	5.6404	5.6350	5.8446

9. Using $x_0 = 1$, $y_0 = 2$, and $h = 0.1$ we have

$$x_1 = x_0 + h(x_0 + y_0) = 1 + 0.1(1 + 2) = 1.3$$

$$y_1 = y_0 + h(x_0 - y_0) = 2 + 0.1(1 - 2) = 1.9$$

and

$$x_2 = x_1 + h(x_1 + y_1) = 1.3 + 0.1(1.3 + 1.9) = 1.62$$

$$y_2 = y_1 + h(x_1 - y_1) = 1.9 + 0.1(1.3 - 1.9) = 1.84.$$

Thus, $x(0.2) \approx 1.62$ and $y(0.2) \approx 1.84$.

Appendix

Appendix I

3. If $t = x^3$, then $dt = 3x^2\,dx$ and $x^4\,dx = \frac{1}{3}t^{2/3}\,dt$. Now

$$\int_0^\infty x^4 e^{-x^3}\,dx = \int_0^\infty \frac{1}{3}t^{2/3}e^{-t}\,dt = \frac{1}{3}\int_0^\infty t^{2/3}e^{-t}\,dt$$

$$= \frac{1}{3}\Gamma\left(\frac{5}{3}\right) = \frac{1}{3}(0.89) \approx 0.297.$$

6. For $x > 0$

$$\Gamma(x+1) = \int_0^\infty t^x e^{-t}\,dt$$

$$\begin{array}{|ll|} \hline u = t^x & dv = e^{-t}\,dt \\ du = xt^{x-1}\,dt & v = -e^{-t} \\ \hline \end{array}$$

$$= -t^x e^{-t}\Big|_0^\infty - \int_0^\infty xt^{x-1}(-e^{-t})\,dt$$

$$= x\int_0^\infty t^{x-1}e^{-t}\,dt = x\Gamma(x).$$

Appendix III

3. Expanding by the first row gives

$$\begin{vmatrix} 2 & 0 & 5 \\ 0 & 7 & 9 \\ -6 & 1 & 4 \end{vmatrix} = 2\begin{vmatrix} 7 & 9 \\ 1 & 4 \end{vmatrix} + 5\begin{vmatrix} 0 & 7 \\ -6 & 1 \end{vmatrix} = 2(28 - 9) + 5(0 + 42) = 248.$$

6. Expanding by the third row gives

$$\begin{vmatrix} 1 & 0 & 9 & 0 & 3 \\ 2 & 1 & 7 & 0 & 0 \\ 0 & 0 & 2 & 0 & 0 \\ -1 & 1 & 5 & 2 & 2 \\ 2 & 2 & 8 & 1 & 1 \end{vmatrix} = 2\begin{vmatrix} 1 & 0 & 0 & 3 \\ 2 & 1 & 0 & 0 \\ -1 & 1 & 2 & 2 \\ 2 & 2 & 1 & 1 \end{vmatrix} = 2\left(1\begin{vmatrix} 1 & 0 & 0 \\ 1 & 2 & 2 \\ 2 & 1 & 1 \end{vmatrix} - 3\begin{vmatrix} 2 & 1 & 0 \\ -1 & 1 & 2 \\ 2 & 2 & 1 \end{vmatrix}\right)$$

$$= 2(1)\begin{vmatrix} 2 & 2 \\ 1 & 1 \end{vmatrix} + 2(-3)\left(2\begin{vmatrix} 1 & 2 \\ 2 & 1 \end{vmatrix} - 1\begin{vmatrix} -1 & 2 \\ 2 & 1 \end{vmatrix}\right)$$

$$= 2(1)(2 - 2) + 2(-3)[2(1 - 4) - (-1 - 4)] = 6.$$

9. We first compute

$$\begin{vmatrix} 2 & 1 \\ 3 & 2 \end{vmatrix} = 1.$$

Then

$$x = \frac{1}{1}\begin{vmatrix} 1 & 1 \\ -2 & 2 \end{vmatrix} = 4 \quad \text{and} \quad y = \frac{1}{1}\begin{vmatrix} 2 & 1 \\ 3 & -2 \end{vmatrix} = -7.$$

12. We first compute

$$\begin{vmatrix} 4 & 3 & 2 \\ -1 & 0 & 2 \\ 3 & 2 & 1 \end{vmatrix} = 1.$$

Then

$$x = \begin{vmatrix} 8 & 3 & 2 \\ 12 & 0 & 2 \\ 3 & 2 & 1 \end{vmatrix} = -2, \qquad y = \begin{vmatrix} 4 & 8 & 2 \\ -1 & 12 & 2 \\ 3 & 3 & 1 \end{vmatrix} = 2, \quad \text{and} \quad z = \begin{vmatrix} 4 & 3 & 8 \\ -1 & 0 & 12 \\ 3 & 2 & 3 \end{vmatrix} = 5.$$

Appendix IV

3. $2z_1 - 3z_2 = 2(2 - i) - 3(5 + 3i) = -11 - 11i$

6. $\overline{z}_1(i + z_2) = (2 + i)(5 + 4i) = 10 + 13i + 4i^2 = 6 + 13i$

9. $\dfrac{1}{z_2} = \dfrac{1}{5 + 3i}\dfrac{5 - 3i}{5 - 3i} = \dfrac{5 - 3i}{25 + 9} = \dfrac{5}{34} - \dfrac{3}{34}i$

12. The modulus is $r = \sqrt{0^2 + 4^2} = 4$ and $\theta = -\pi/2$, so

$$z = 4e^{-i\pi/2}.$$

15. The modulus is $r = \sqrt{2^2 + 2^2} = 2\sqrt{2}$ and $\theta = \pi/4$, so

$$z = 2\sqrt{2}e^{i\pi/4}.$$

18. The modulus is $r = \sqrt{10^2(3) + 10^2} = 20$ and $\tan\theta = 10/(-10\sqrt{3}) = -\sqrt{3}/3$, where θ is in the second quadrant, so $\theta = 5\pi/6$ and

$$z = 20e^{5i\pi/6}.$$

21. $z = 8e^{-i\pi} = 8[\cos(-\pi) + i\sin(-\pi)] = -8$

24. We first express $1 + i$ in polar form as

$$1 + i = \sqrt{2}\left(\cos\frac{\pi}{4} + i\sin\frac{\pi}{4}\right).$$

Then

$$(1 + i)^{10} = (\sqrt{2})^{10}\left(\cos\frac{10\pi}{4} + i\sin\frac{10\pi}{4}\right) = 32\left(\cos\frac{5\pi}{2} + i\sin\frac{5\pi}{2}\right) = 32(0 + i) = 32i.$$